Magic & Rationality
The magical logic of manifesting success

Draft 2.3-RC-13
~~Private and confidential~~
Michel Gagnon

Zip Ditto

© July 17, 2022

Contents

Contents		3
Preface		7
Outline		9
I	**Magicians & Skeptics**	**11**
1	The supposed secret to success	13
2	The collaboration game	17
3	ESP	23
4	Zen ESP	29
5	Magician's ESP	33
6	ESP's greatest magician	41
7	Rationality	45
8	Deeply believing in success	51
9	Magician's rationality	57
10	Caveats to magician's rationality	63

11	The Zen lemonade garden	69
12	Lemons & lemonade	81
13	Thoughts for skeptics	91
14	Thoughts for magicians	99
15	The wish giving gem	107

II	Appendix: Magician's rationality	111
16	Rationality	113
17	Superrationality	117
18	Hyperrationality	121
19	Misc	123
20	Introduction	125
21	Sufficiently logical humans	127
22	Belief-wise games	133
23	Theory of magician's rationality	139

III	Appendix: Zen lemonade garden	153
24	Introduction	155
25	Model	159
26	Balancing teaching and studying	163

CONTENTS 5

27 Genuine quitters **169**

28 Sybil attacks **171**

29 Conclusion **175**

CONTENTS

Preface

Let me tell you a story. A long time ago, I used to conduct science experiments and write math proofs for a living. I worked for MIT, studied at Harvard, and consulted to DARPA. I was a professional skeptic—a computer scientist specializing in security.

And then my career and my life swiftly crashed one day. I had been living under the theory, the life philosophy that if I acquired everything I ever wanted, and if I accomplished everything I ever wanted to accomplish, I would be happy. I had a lovely wife, a lovely dog, a lovely house, lovely friends, a lovely family, a lovely job, lots of money, and I had accomplished everything I had set out to accomplish.

I should have been happy, I thought to myself, and yet, I realized one day I was profoundly miserable. My life philosophy must be fundamentally flawed, I thought—but I didn't know where the flaw was. So, I took a drastic measure. To uproot the root, I decided to become skeptical of skepticism.

In rapid time, I replaced skepticism with a belief in magic. I had many adventures, and slowly, over the course of nine years, bit by bit, I reacquainted myself with skepticism and rigorous inquiry. And yet today, my faith in magic remains. This book represents a fruition of my journey.

I have taken some of the magical lessons I have learned and subjected them to logical analysis and simulation experiments. In particular, I have investigated the "supposed secret to success"—the theory that *if you believe you will succeed, then you will succeed*. The results of my investigation surprised even me.

I set out to write a book for magicians and skeptics, a book about magic and math. When I began writing, I didn't have any math proofs—just aspiration. I explored possibility after possibility, searching for an interesting thing to prove about magic. For every math proof in this book, there are probably ten pages of math in the trash.

Then, shortly into my investigation, a realization popped into my head: magicians like to collaborate with each other, because they believe in each other. I sketched out a proof, and it made me smile and laugh.

I went down the rabbit hole, exploring the consequences of that realization. The theorems, proofs, conjectures, and experiments just rolled out. It took a month to complete the first draft, and now, six months later, I believe I have a fascinating story for you. I hope you enjoy it and find it enlightening.

Outline

This book uses logic, philosophy, and whimsical storytelling to investigate the magic of succeeding by believing.

In particular, we look at the "supposed" secret to success: *if you believe you will succeed, then you will succeed.* What happens when you adopt this belief? The logical consequences may be surprising. For example, under certain conditions, this supposed secret becomes a logical, self-fulfilling prophecy.

Through a progression of 15 chapters, we follow a narrative—going deeper and deeper into the rabbit hole of the supposed secret. Most chapters include a short story to illustrate a logical concept relating to the supposed secret, and most chapters also include a rigorous analysis to satisfy the skeptics. My fundamental thesis is that "magicians" (those who believe in the supposed secret) are logically empowered by the supposed secret to success—even according to skeptical logic.

Our narrative begins in Chapters 1 and 2, with the realization that magicians like to collaborate with each other, because they perceive each other as successful. However, skeptics might avoid magicians because skeptics might perceive magicians as having unwarranted self confidence.

Thus, if magicians wish to collaborate with skeptics, too, then they need a way to hide their magic from skeptics, while revealing it to their fellow magicians. In Chapters 3 – 5, "ESP" comes to the rescue—a form of covert communication.

But there's a problem with ESP: less advanced magicians have difficulty identifying more advanced magicians. To remedy this issue, Chapter 6 presents a method for magicians to work together to identify more advanced magicians.

Building towards our logical crescendo, we study *magician's rationality* in Chapters 7 – 10. Magician's rationality shows that the supposed secret can be a self-fulfilling prophecy. The gist is that magicians are logically obligated to help each other out (or at least not thwart each other), because they believe each other will be successful. To betray a fellow magician is to betray your own personal belief in their success. There are caveats to this obligation, so it only works under certain conditions.

We reach the logical crescendo of this book in Chapters 11 & 12. It may seem that magician's rationality could actually be a burden. For, if you are logically obligated to help your fellow magicians, then what would happen if you were bombarded with requests for help? Worse still, in a light-hearted worst-case scenario, what if an army of undercover skeptics were asking for help too, trying to waste your time? How could you possibly help everyone while also pursuing your personal interests at the same time? I present *Zen lemonade garden* (ZLG) as the solution. ZLG prescribes transmuting lemons into lemonade. For ZLG to succeed, magicians must not betray each other. Thus magician's rationality comes into play. The Zen lemonade garden is inspired by research that I began at MIT, Harvard, and DARPA, then simmered in Buddhism for 9 years.

Next, I present two chapters: "Thoughts for skeptics" and "Thoughts for magicians." The first points skeptics towards magic, and the second points magicians towards skepticism. This book concludes with a short story about a little girl who is haunted by uncountable wishes.

Lastly, I include 67 pages of appendices, for interested skeptics. The appendices include a detailed formal proof of magician's rationality, and numerical results from simulating the Zen lemonade garden on my computer.

Part I

Magicians & Skeptics

Chapter 1

The supposed secret to success

> If you believe you will succeed,
> then you will succeed.

Thus goes the "supposed" secret to success. I put the word supposed in quotes, because this book is meant for both believers and nonbelievers—*magicians* and *skeptics*.

1.1 Magic

The supposed secret to success is a "magical" belief, so let us call the believers *magicians*.

But, what really do I mean by magic? Opinions differ between skeptics and magicians. But, as a mathemagician, I define magic simply as *mysterious impressive phenomena*.

Magic is a matter of perspective. What might be *mysterious* to one person, might be very well understood by someone else. Also, what might be *impressive* to one person, might be dismissed as a fairytale by another.

1.2 Skepticism

Skepticism is doubt. It is a stance of suspicion. As Carl Sagan put it, "Extraordinary claims require extraordinary evidence."

The supposed secret to success is an extraordinary claim. Can you really succeed, just by believing? It seems outrageous. Therefore, skeptics are skeptical.

1.3 The tension between magic and skepticism

There of course exists a tension between magic and skepticism. Magical beliefs, such as the supposed secret to success, defy skeptical perspectives. Magic must often be taken on faith or belief. In the case of the supposed secret to success, the consensus of science declares we lack extraordinary evidence.

Furthermore, there also exists a tension between the believers and nonbelievers—the magicians and skeptics. Skeptics often dismiss the believers, and also the other way around. Ideologically, we tend not to get along.

But, as an ambassador of both magic and skepticism, I am pleased to present this book to both sides. In this book, I do not attempt to prove that the supposed secret really, truly works, in general. However, it is clear that if you choose to believe in the supposed secret, there are consequences. For example, skeptics might dismiss you.

This book explores these consequences using my skeptical toolkit, namely math, formal logic, and computer simulation. I hope each chapter illuminates the possibilities and fascinates you as well.

1.4 Fundamental thesis

Perhaps the supposed secret to success actually, truly works. Or maybe the believers, the magicians, maybe we're all mis-

1.4. FUNDAMENTAL THESIS

taken. Regardless of the truth of the supposed secret to success, my fundamental thesis is that the supposed secret to success can plausibly empower magicians to accomplish surprising feats.

How do I demonstrate that such surprising feats are plausible? I use math, logic, and simulation from a mostly neutral perspective. As in, for most of my analyses, I neither presume the supposed secret is true, nor false. Rather, I merely presume certain people believe it, while others do not. The general pattern I discovered is that there are surprising consequences to believing in the supposed secret to success.

For example, under certain conditions, becoming a magician can increase your overall quantity of collaborations—which takes us to the next chapter.

Chapter 2

The collaboration game

Let's do a thought experiment. Specifically, let's begin by hypothetically supposing you are a struggling musician, but have just now began believing in the supposed secret to success.

Congratulations! You are now a magician. And, you have just joined a magical group of people on Earth who believe *if you believe you will succeed, then you will succceed*. Because this is your belief, you should now begin believing in your success, and thereby, now, you will succeed (at least, so you believe). What will your struggles look like now?

For the next step, suppose you bump into a guitarist named Alice at a coffee shop, strumming for fun on the back patio. After some chit chat, you both discover each other are magicians. Great!

Here's where it starts to get interesting. The supposed secret to success is available to everyone, or at least that's what the believers believe. As in, *if someone believes they will succeed, then they will succeed*. Therefore, you believe Alice will succeed (since you believe she's a magician), and she believes in you, too (since she believes you're a magician).

So, the question then arises, should you two collaborate? Well, the chances of collaboration have just gone up, since you both perceive each other as destined for success!

Thus, we can see that becoming a magician can help you

to magnetize collaborations with your fellow magicians, since magicians perceive each other as destined for success. Therefore, if you want to increase your quantity of collaborations, just choose to believe. But wait, there's a hiccup.

2.1 Hiccup

Unfortunately, while becoming a magician might magnetize collaborations with magicians, becoming a magician might repel collaborations with *skeptics*—nonbelievers. The issue is: skeptics might be suspicious of magicians, because magicians have "unwarranted" self confidence (unwarranted from the skeptics' perspective).

If skeptics can spot magicians, then becoming a magician might lower your overall quantity of collaborations—especially if skeptics outnumber magicians. The next chapter presents an approach magicians can use to cope with this hiccup.

But before we discuss coping with this hiccup, I would like to dedicate the rest of this chapter to my mathematical "model" of the collaboration game. We begin with a discussion of *models*, in general. What is a model? How are they useful? Can we trust models?

2.2 Models

Models are simplifications of reality. My story above presents a model of collaboration. According to my above model, you can increase your quantity of collaborations simply by becoming a magician. Unfortunately, I believe this model is too simple. For the model omits an important possibility: skeptics might avoid magicians, which could decrease your collaborations, overall.

In general, modeling is an art. We want to simplify reality, so we can analyze it. But if we simplify reality too much, then we miss crucial details. And if we miss crucial details, our analyses could be misleading.

So, how can you tell if a model is good? How can you tell if a model is trustworthy? Is a model too simple? Is it missing important details?

Just as modeling is an art, judging models is also an art. Judgement, in this case, requires a sense of what's really going on in reality—a sense of what is relevant and what is not, and what is possible and what is not.

I say all this, because I have crafted many models for this book. With the exception of the collaboration game, I claim each of my models is "good." By that I mean: I claim my models capture reality.

However, it is up to you and your understanding of reality, to judge my models for yourself.

2.3 Mathematical models

There are roughly two types of models I employ in this book: (1) *informal* models, and (2) *mathematical* models. Informal models are simply stories told in English, while mathematical models are stories expressed in the language of mathematics. For this book, mathematical models are beneficial for one key reason.

Specifically, mathematical models empower skeptics to conclusively analyze my stories. If my mathematical models and analyses are reasonable, then there should be no debate: skeptics ought to agree with my claims in this book.

The primary downside to mathematical models is that they are obfuscated and unintelligible to mathematical outsiders. Thus, I do not exclusively rely upon mathematical models. Rather, I always include a matching informal story for every mathematical model in this book. The informal stories tend to omit details for the sake of good storytelling. My intention though, is for my stories is to faithfully illustrate their accompanying mathematical model.

With this understanding of mathematical models in mind, I present this book's first mathematical model and analysis. However, do not be fooled by the charm of the fancy symbols that proceed. This mathematical model misses the hiccup of Section 2.1; this model fails to take into account the possibility that skeptics avoid magicians. Nevertheless, I include this proof and model as a case study in the fallibility of mathematical modeling.

2.4 Mathematical model for the collaboration game

Definition 1. *The collaboration game*

Two rational players, P_1 and P_2, must decide if they want to collaborate with each other. Each player votes. If both players vote to collaborate, then they collaborate, otherwise they do not. The objective of the game, for each player, is to increase their chances of collaborating. The game is made difficult because each player is burdened by a risk-mitigation formula that prevents voting for players who seem sufficiently unlikely to "succeed." Each player's only choice is to decide whether or not to believe in the supposed secret to success.

Definition 2. *Player information*

Let s_i be the event that P_i "succeeds."

Let x_i be a mutable boolean variable that indicates P_i believes they will "succeed." The value x_i is initialized randomly according to a Bernoulli distribution, and the only way x_i can be updated is by P_i choosing to set $x_i = true$ as a consequence of believing in the supposed secret to success (Conjecture 1).

Let y_i be an immutable boolean value that indicates P_i believes in the secret to success (formalized in the discussion of Conjecture 1). The value y_i is chosen randomly according to a Bernoulli distribution.

Let $U_{i,j}$ = the universe of information P_i knows about themselves and also P_j. Informally: players do not know the values for s_i. Player i knows their own x_i value. Player i knows y_i and y_j. Formally: $U_{i,j} = (x_i, y_i, y_j)$.

2.4. MATHEMATICAL MODEL FOR THE COLLABORATION GAME

Conjecture 1. $\forall k.P(s_k|x_k = \text{true}) \geq P(s_k|x_k = \text{true} \vee x_k = \text{false})$

Conjecture 1 represents a milder version of the supposed secret to success. In plain English, it says: if you know someone believes in their own success, the probability that they succeed increases. Of course if the supposed secret to success is true, then this milder conjecture is also true.

Notation 1. *Perceived probability*

Let's use the following notation. $\mathbb{P}_i(a)$ represents the probability of a according to Player i's perception.

Definition 3. y_i

The value y_i indicates whether or not P_i believes in the secret to success. Formally: $(y_i = \text{true}) \implies \forall k.\mathbb{P}_i(s_k|x_k = \text{true}) \geq \mathbb{P}_i(s_k|x_k = \text{true} \vee x_k = \text{false})$.

Definition 4. *Collaboration*

Let $c_{i,j} = c_{j,i} =$ the event that P_i and P_j collaborate. P_i and P_j collaborate if, and only if P_i votes for P_j, and vice versa. Let $v_{i,j} =$ the event that P_i votes for P_j. Of course, $P(c_{i,j}|v_{i,j} \wedge v_{j,i}) = 1$, otherwise $P(c_{i,j}) = 0$.

Lemma 1. $P(c_{i,j}|v_{i,j} \wedge v_{j,i}) = P(v_{i,j}) \times P(v_{j,i})$

Proof. Straightforward consequence of the Definition 4. □

Definition 5. *Risk-mitigation formula*

Player i votes for P_j if, and only if P_j seems sufficiently likely to succeed. More formally, as the perceived probability of success goes up (for the other player), so does the probability of a voting for that player. Formally, $\mathbb{P}_i(s_j|a = \text{true}) \geq \mathbb{P}_i(s_j|a = \text{true} \vee a = \text{false}) \implies P(v_{i,j}|a = \text{true}) \geq P(v_{i,j}|a = \text{true} \vee a = \text{false})$, for all a.

Theorem 1. $P(c_{i,j}|y_i = \text{true}) > P(c_{i,j}|y_i = \text{false})$

Proof. Assume you, P_i, believe in the secret to success and the other player believes too. I.e. $y_i = y_j$ = true.

Therefore, according to Definition 3, you both believe Conjecture 1, i.e. $\forall k.\mathbb{P}_h(s_k|x_k = \text{true}) \geq \mathbb{P}_h(s_k|x_k = \text{true} \vee x_k = \text{false})$ for $h = i$ and $h = j$. Consequently, you both believe setting x = true is a good idea (conducive to success). Formally: $\mathbb{P}_h(s_g|x_g = \text{true}) \geq \mathbb{P}_h(s_g|x_g = \text{true} \vee x_g = \text{false})$ for all h and g. Since you're both rational, you each set $x_i = x_j$ = true.

And because y_i and $y_j \in U_{i,j}$ and because y_i and $y_j \in U_{j,i}$, you both know that each other believe in the secret to success. Furthermore, since you both know you're both rational, you both know $x_i = x_j = true$. Since you believe in the secret to success (i.e. y_i = true), and you know $x_j = true$, then the secret to success implies $\mathbb{P}_i(s_j|x_j = \text{true}) \geq \mathbb{P}_i(s_j|x_j = \text{true} \vee x_j = \text{false})$, and similarly for Player j.

The risk-mitigation formula now comes into play. Recall: $\mathbb{P}_i(s_j|a = \text{true}) \geq \mathbb{P}_i(s_j|a = \text{true} \vee a = \text{false}) \implies P(v_{i,j}|a = \text{true}) \geq P(v_{i,j}|a = \text{true} \vee a = \text{false})$, for all a. Thus, the implication applies when we let $a = x_j$ and we know $P(v_{i,j}|x_j = \text{true}) \geq P(v_{i,j}|x_j = \text{true} \vee x_j = \text{false})$. And similarly for P_j.

Since we know $x_i = x_j$ = true, then the probability of voting for each other is increased.

Finally, by Lemma 1, we know that the probability of collaboration has just increased. Thus we have established that when two players believe in the secret to success, their chances of collaborating increase. We conclude by wrapping up our case analysis. Suppose neither player believes in the secret to success. Then, it's clear they can't take advantage of the secret to success to boost their perceived chances of success. But, what if just one player believes in the secret to success? Well, the magician would not perceive the skeptic to be boosting their own chance of success, so the magician would not be more likely to vote for the skeptic. And since the skeptic doesn't believe in the secret to success, the skeptic isn't impressed by the magician's belief system, so the skeptic would not be more likely to vote for the magician. Thus the chance of collaboration does not increase when a magician and a skeptic pair up. □

Chapter 3

ESP

3.1 Review

In Chapter 1, to quickly review, I defined a magician as someone who believes in the supposed secret to success:

*If you believe you will succeed,
then you will succeed.*

Then, in Chapter 2, I argued that magicians like to collaborate with their fellow magicians, because they perceive each other as successful. Thus, if you join Team Magic, you might increase your total quantity of collaborations.

Unfortunately, there's a hiccup with that theory. Essentially, skeptics might be wary of collaborating with magicians, thanks to all our "unwarranted" self confidence.

So, on one hand, joining Team Magic might increase your collaboration with magicians. However, on the other hand, skeptics might stay away. Thus, becoming a magician might hurt your overall prospects at collaboration, especially if skeptics outnumber magicians.

3.2 Introduction

It would be nice if we, magicians, could keep our magical status secret, because then we could collaborate with skeptics (who might otherwise discriminate against us). At the same time, we would like to spot our fellow magicians, since we especially want to collaborate with each other.

The question is: how do we balance the two? How can we keep our secret from skeptics, while at the same time spotting the magic of our fellow magicians?

What if, just what if magicians had a 6th sense for detecting magicians, a type of clairvoyance, a type of *ESP?* I believe such ESP is possible, and describe the general idea of ESP in this chapter.

ESP (the way I define it) is simply when you can guess someone holds a particular belief, because you saw them behave a particular way. For example, suppose you witness someone step into a Catholic church. Even in the absence of other clues, there's a decent chance that person is Catholic.

Such ESP happens intuitively everyday, but what we're really interested in is *esoteric ESP*. The premise is that when we behave according to our *esoteric beliefs*, then, at times, *only* our fellow believers will be able to deduce our beliefs from our behaviors. As in magicians have an advantage over skeptics at spotting magicians.

Let us begin with a fable about esoteric ESP that takes place during a chess match in 1925.

3.3 The legend of crazy king

According to legend (this is how I begin fables), in 1925, in Zürich, two of the greatest chess players of all time met for a chess match, in a large sold-out theater. A Russian versus a Swede. Towards the beginning of the middle of the game, the Russian picked up his King, moved it up, floated it right into the center of the board, set his King down on a new square, right there in the center, let go and immediately slapped his forehead and shouted *Doh!*

The entire crowd gasped. The Russian had made a blunder! No grandmaster would move their king to the center of the board. Well, almost everyone gasped, for there was a Brit in the audience who grinned.

The Brit knew better, because she had been practicing the esoteric *crazy king* strategy against herself, all on her own, for the past ten years. The Brit knew the Russian was playing this esoteric strategy.

3.4 Analysis of the legend

In the legend, the Russian and the Brit both believed in the esoteric crazy king strategy. As a consequence, when the Russian made his move, the Brit was able to infer that the Russian believed in the esoteric strategy. At the same time, everyone else in the room was clueless. This inference is an example of *esoteric ESP—inferring another's esoteric beliefs based upon observations of their esoteric behavior.*

There's a problem with crazy-king ESP, though. Fast forwarding to today, you can study the crazy king strategy in books (according to legend). These days, any time an advanced chess player thrusts their king into the ruckus, you can be confident they're playing crazy king. The secret is out. Everyone spots it now, even when they haven't labored to master the crazy king strategy themselves.

The secret of crazy king ESP has been *spoiled*; crazy king is now easily be spotted by everyone.

How then do we achieve a version of esoteric ESP that never spoils and always remains fresh? And, can magicians use ESP to identify their fellow magicians, while remaining incognito amongst the skeptics? In other words, does ESP apply to the supposed secret to success? The next chapter is dedicated to these questions.

But, for now, to conclude this chapter, allow me to present to you the *magician's chess strategy*—a strategy magicians can use to covertly identify each other while playing chess.

3.5 Magician's chess strategy

To demonstrate esoteric ESP via a computer simulation, I have devised a chess strategy, the *magician's chess strategy*. Magicians can use this strategy to secretly identify each other while playing chess. And, as the simulation results show, artificial intelligence can use the strategy, too.

The magician's strategy has the following properties: (1) it is optimal with respect to your skill level, (2) it is easy to explain, (3) it takes significant time and effort to learn, and (4) once learned it takes little effort to apply.

The strategy is esoteric—understood by few. And when I begin explaining the strategy, in the next sentence, you will see why it is understood by few. To learn the strategy, begin by memorizing the following string of letter-number pairs:

```
d4 d5 c4 c1 a6 f8 a7 b5 h4 b8 c3 d1 c6 h6 e6
e7 f4 e2 a5 e1 h3 c5 d7 e8 g3 f5 b2 g7 g5 a1
d2 h1 f2 b6 g4 g8 a2 d8 h5 b4 h2 a4 e3 c2 f1
f3 g6 c8 e4 b1 h8 g2 c7 a8 h7 e5 a3 b7 f7 d6
b3 g1 f6 d3
```

The string encodes the location of every square on the chess board. For example, the string begins with d4, which represents the square three spaces ahead of where the white queen starts.

You use the strategy in the game as follows. Let's say it's your turn, you're searching for the best move, you've narrowed down your search to two or more different moves, but you're stuck—they all seem equally good, bad, or neutral. Allow the magician's strategy to break the tie.

Here's how. Go through the string, in sequence, and stop immediately once you've reached a square where one of your moves may land. Then, perform that move. In the unusual case where multiple pieces may move to that square, then break that tie by going through the string (once again), but this time stopping when you reach a square that currently holds your tied piece. That's it. Make sure to practice.

3.5. MAGICIAN'S CHESS STRATEGY

For example, if you're playing white, and you can't decide how to start, then move your pawn to d4.

Unfortunately for you, the magician's strategy has limitations when used in the real world. I can think of at least three. First, what are the chances you will bump into a magician who has also spent the time learning this strategy? Second, there are skeptics with photographic memories. Third, to be able to detect a magician via this form of ESP, you will need to know when they can't make up their mind and are leaning on the magician's strategy for a tie breaker.

Regarding the third limitation, I met someone this morning, and we rather quickly discovered we are both interested in spy movies. It also turns out he is an avid chess player—he can look ahead seven moves, in his head, when pondering the best move to make. I chatted him up, and he told me he can detect when a less advanced player can't make up their mind.[1] Since he finds spy movies interesting, I told him about magician's chess. His response: esoteric ESP via the magician's strategy is plausible. Cool, right?

In case you're interested, the magician's strategy is based on the concept of *side channels* from the field of computer security. The gist of the concept is there's a "primary channel" which is supposed to convey information, and then there are "side channels" which might contain information, but people tend not to look there for information.

OK, let's get into a computer-science experiment.

---

[1] Here's how. First, he precisely estimates his opponent's skill level. How? He waits until his opponent makes their first mistake. This suggests the player can look ahead six moves or fewer. So, now he wants to know: can my opponent lookahead six moves? Or is it fewer? So, he deliberately plays a move that results in an optimal depth-six counter move, and a suboptimal depth-5-or-fewer counter move. If the opponent makes the suboptimal move, then he has approximately ruled out the possibility that the player can play to depth 6. And so on, until he finally figures out how good his opponent is. Cool, right? From there it's relatively straightforward to estimate when the opponent can't make up their mind.

3.6 Computer science experiment

To do

Chapter 4

Zen ESP

4.1 Review

To collaborate with skeptics, magicians might want to keep their magical beliefs a secret, since skeptics might discriminate against magicians. At the same time, magicians want to meet other magicians—because, magicians especially appreciate collaborating with each other. So, how do magicians go about hiding and sharing their magical beliefs? There seems to be a chance "ESP" can help.

In the last chapter, I presented my general concept of *esoteric ESP*. In a nutshell, esoteric ESP is when you can infer someone holds an esoteric belief, by observing their esoteric behavior. Furthermore, at times, only believers will be able to make such inferences. Thus, ESP offers a form of covert communication.

With regards to the supposed secret to success, ESP might be useful for magicians. Perhaps magicians could detect fellow magicians, while skeptic eavesdroppers would remain clueless.

If in fact magicians could use ESP this way, then it would solve a major problem for them. Specifically, magicians could meet other magicians, without revealing their magical beliefs to skeptics. Magicians could collaborate with everyone!

4.2 Introduction

Before discussing ESP applied to the supposed secret to success, allow me to tell you a true story of *Zen ESP* from last Sunday. My goal here is merely to demonstrate the plausibility of esoteric ESP in everyday life, in the real world.

In this story, a Zen insider was able to spot my Zen, while I'm pretty sure most everyone else was clueless. It takes place inside a coffee house named ESP.

4.3 ESP, the coffee house

There's a coffee shop in my neighborhood called ESP, and the patrons here tend to be magicians. Last Sunday, at open-mic night, I wanted to impress someone in the crowd named ▇.

So, I grabbed the mic to read a riveting story, a draft of the *legend of the Zen lemonade garden*.[1]

Being a magician, I chose to believe I would succeed. However, I bombed: no one in the crowd followed my story. Is this proof that the supposed secret to success isn't true? Keep reading.

When I realized I had bombed, I said, "Oops," hung up the mic, and began to step off stage. But, a gentleman named ▇ shouted, "Don't give up!" and a gentleman named ▇ shouted, "You got it!"

So, I smiled, picked up the mic, and succeeded at a brief bit of improv comedy. ▇ laughed, and I hung up the mic with a smile.

Now, to get to the point. Later that night, a stranger came up to me, and said, "My dude, you're Zen."

I was surprised anyone would recognize me as Zen after my bomb, so I asked "You followed my story?"

They laughed, and said, "Nope!"

"Then how do you know that I know Zen?"

They said, "Bruv, you've got Zen confidence."

---

[1] See Chapter 11.

4.4 Zen confidence

Only one person seemed to recognize the Zen of my confidence, the Zen of my rebound. Why only one? To answer, let's dig into a little backstory on Zen confidence, and how it relates to the supposed secret to success.

I've been studying and practicing Buddhism for years, according to many books, among them a book called *Zen and The Art of Happiness*, by Chris Prentiss. According to this book, at least if you read between the lines, you should magically manifest success by finding ways to view personal failures as personal successes.

In the case of open-mic performances, I do not regard bombing as a failure; rather, bombing is a lesson. Thus, no matter what happens at open mic night, I either succeed at the performance, or I succeed at learning a lesson. Thus success is guaranteed.

Furthermore, I speculate this approach gives me confidence, for I am less worried about "failure," and this confidence actually helps me not bomb—sometimes.

4.5 Zen ESP

Did *you* know confidence is a part of Zen? Are you familiar with Zen? Do you believe in Zen? Are your answers, "yes, yes, yes?" Or, are your answers "no, no, no?"

This story illustrates the potential for esoteric ESP in everyday life. In the case of this story, only someone who was familiar with Zen was able to identify me as Zen, via my confident behavior. Furthermore, I suspect only Zen insiders tend to be aware that confidence is a part of Zen.

Lastly, confidence seems to be a part of the supposed secret, too. After all, if you know you're destined to succeed, wouldn't you be confident? Which, brings us to our next chapter, *Magician's ESP*.

Chapter 5

Magician's ESP

5.1 Review

In Chapter 2, we saw saw that skeptics might be wary of the "unwarranted" self confidence of magicians. Thus, magicians might wish to remain incognito, and only share their magical status with their fellow insiders.

Chapter 3 introduces ESP and esoteric ESP. ESP is simply when you can guess someone holds a particular belief, because you saw them behave a particular way. Esoteric ESP is when you can spot esoteric beliefs, simply by observing their esoteric behavior. A key aspect of esoteric ESP, is insiders have an advantage at spotting their fellow insiders.

To demonstrate esoteric ESP in everyday life, Chapter 4 tells a story about how a Zen insider spotted a fellow Zen insider, using esoteric ESP. Essentially, confidence is a part of Zen, and if you witness confidence (mixed together with other clues), then you might be able to spot Zen. But only if you're familiar enough with Zen to connect the dots.

5.2 Introduction

I imagine it's unsurprising that Zen ESP is possible. After all, Zen represents an entire religion and seemingly a huge body of beliefs and teachings. Would it really be surprising that insiders can perform esoteric ESP, since there is such a vast body of beliefs and behaviors that unite them?

So, the question of this chapter is, can magicians employ esoteric ESP—even though there is only one belief that unites us? Can magicians use esoteric ESP to identify each other without skeptics noticing? My answer to this question is: *plausibly*. I personally believe in magician's ESP, and I believe I can demonstrate to you that magician's ESP is at least plausible.

My central argument is that if you believe in the supposed secret, you will end up acquiring many other accompanying beliefs as well. Thus the supposed secret to success is like the trunk of a bushy tree, and the myriad branches and leaves are the many lessons to live by. Therefore, the supposed secret to success is like Zen in an important regard: there are many lessons to live by, which helps you become spotted by other believers, and helps you spot other believers. Lastly, since magicians will tend to be more familiar with these beliefs, then skeptics will be less able to spot magicians, compared to magicians.

To demonstrate the plausibility of the bushy tree of the supposed secret, this chapter offers anecdotes—lessons I have learned from the supposed secret. Also, I discuss how these lessons affect behavior, and thus how fellow believers can spot each other. Lastly, I explain why these lessons are plausibly universal—not just unique to me. Therefore, there is hope magicians can spot magicians, in general.

5.3 Confidence

If you actually deeply believe in the supposed secret to success, you ought to be confident, right? The logic is simple: you will be successful, and that is pretty much the definition of confidence.

Therefore, if you witness confidence in someone, then that could be a clue to them being a magician. But what does their confidence look like? How does it smell? Does the confidence reveal the swagger of a neophyte believer, or does it contain the subtle notes of a grandmaster magician effortlessly weaving their way of life? Or, it could be the confidence of an expert craftsperson, unrelated to magic. Or, something else.

There are many flavors of confidence, and I believe certain flavors are more associated with magic than others. Rather than attempting to pin down an analysis of magical confidence, I'll let you be the appraiser. I'm not sure I'm sufficiently expert on the topic anyway.

But, supposing you catch a whiff of apparent magical confidence, can you then be sure you're seeing a magician? No, of course not. In general, you need to combine clues together to identify magicians. Perhaps you could approach and drop a subtle reference to magic. Do they respond with a subtle reference? Can you play off each others references until you both know each other are magicians?

That almost concludes our discussion of confidence, with regards to magician's ESP. I want to bring up two more items though. First, it is probably uncommon to develop confidence overnight, simply by willing yourself to believe in the supposed secret to success. Rather, I expect it is more common for confidence to develop over time, as your belief in the supposed secret to success deepens. Second, because confidence grows over time (and changes in flavor over time), then the signals emanating from more advanced magicians will differ from the signals emanating from more novice magicians.

5.4 Believing in the success of others

Through the trials and tribulations of choosing to believe in the supposed secret to success, I have learned many lessons. I can easily articulate many of them. Other lessons, I can't quite describe. I also believe I've learned unconscious lessons—lessons that affect my behavior, yet I can't enumerate them.

One lesson that I *can* articulate is: *many people tend to underestimate their own likelihood of success.* For people who believe in this lesson, how does it affect our behavior? It increases our interest in collaborating with diamonds in the rough.

I'll tell you a story that took place at ESP, the coffeeshop, last night. I was hanging out with a stand-up comic named David. We were chit chatting, waiting for our turn at the mic.

I asked him if he believed in the supposed secret to success. He said, "Well, I don't believe a belief in success guarantees success, but if you don't believe in yourself, why would anyone else believe in you?"

Following his logic, if people want to collaborate with people who appear successful, and if believing in yourself makes you appear successful, then perhaps believing in yourself makes more people want to collaborate with you. Cool, right?

But the catch is, if you don't believe in yourself, then potential collaborators might stay away. However, and here's the important part, I won't avoid you, because I believe you have *underestimated* yourself—I believe your chances of success are greater than what you perceive, and what other people perceive.

I see a diamond in every rough, therefore I want to collaborate with everyone. And when sufficiently advanced magicians witness my interest in collaborating with rough diamonds, they might guess I am a magician.

Anyway, last night, David had a great set and I bombed. Again. That's two bombs in a row. Having earned my second valuable lesson in two weeks, I believe I will succeed at next week's open mic, when I present this chapter to the crowd.

5.5 Lessons earned over the years

I believe practicing the magic of the supposed secret to success yields incremental fruit. I will share some lessons I have learned over the years, from studying and practicing. First lesson: have fun.

5.5. LESSONS EARNED OVER THE YEARS

I learned to have fun at the beginning, and it has stuck with me. Perhaps it takes others time to learn this lesson, I don't know. But I am sure though, that I do witness much merriment in my fellow magicians. I think having fun is important, because you can have fun a million different ways, and if you apply the supposed secret to success a million different ways, then success will surely float up. Then you can combine your talents together in unique ways.

But for every rule, as the wisdom goes, there's at least one exception. For example, meditation isn't always fun for me, but I do it anyway.

Getting back to my main point, I have learned many lessons, incrementally over the years. I believe my fellow magicians have learned similar lessons, albeit tailored to the individual tastes and personalities of the practitioners. Since I believe the same lessons are available to all of us, we don't necessarily need to read the same books to be able to spot each other's magical behaviors and beliefs. If we just live according to the supposed secret, I believe we will learn many of the same lessons—empowering magician's ESP.

To sum up some lessons I have learned, I have created a diagram containing a *decision tree*. See Figure 5.1. This decision tree symbolizes my approximate thinking process on whether or not I should aim to succeed at something. Bear in mind, for every decision point in the decision tree, there are bound to be exceptions to the rule. Also, this decision tree omits many important considerations, such as risk analysis. It's a simple tree, just to make a point.

It took years to arrive at this decision tree, and I suspect other magicians have similar decision trees in their brains.

You may be noticing a theme from the example lessons I have been sharing: the theme of *time*. Recall, I suspect it usually takes time to cultivate magical confidence. And, it probably takes time to realize others are underestimating themselves. It took years for me to develop the decision tree in Figure 5.1. In general, magic takes time, and thus neophytes will yield different clues than grandmaster magicians, and everything in between.

Figure 5.1: Should I aim to succeed at something?

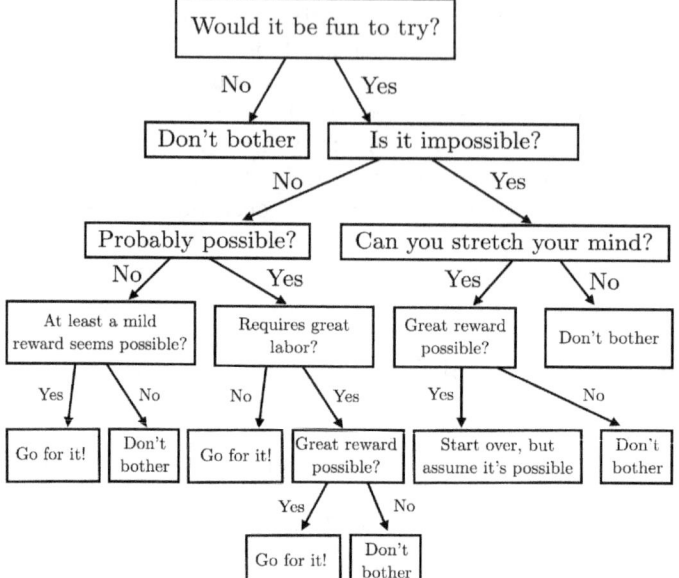

Allow me to elaborate on the time it takes to develop this decision tree. Supposing my decision tree represents actual wisdom, and others have discovered similar wisdom too, then how much time would it take for a neophyte to learn to stretch their mind to attempt the impossible? When to give up? How to know when to give up? When not to give up? What things you should never give up on? What's probably possible? When is great labor warranted? And so on.

Here comes an important point: if you don't intuitively know the answers to those questions, how would you be able to assess whether or not a fellow magician is wise to give up on something? Or whether or not they are wise to attempt the impossible? One of my theses for this chapter is: less advanced practitioners may have difficulty assessing the competency of more advanced practitioners. I theorize that ma-

5.5. LESSONS EARNED OVER THE YEARS

gician's ESP may allow more advanced adepts to identify less advanced adepts, but not necessarily the other way around.

Let's consider a concrete example. Suppose you have just become a believer in the supposed secret to success; you are now a magician. You go to ESP, the coffee shop, and witness a comedian bomb at open mic night. Clearly that person is no magician, for if they were a magician, they would have succeeded, wouldn't they? However, more advanced magicians of course know that "failure" isn't always failure, therefore the comedian might just be an advanced magician after all. Look for more clues.

Which brings us to the next chapter. Given this limitation of magician's ESP (the limitation that less advanced magicians might have difficulty identifying more advanced magicians), how can magicians come together to identify the greatest magician in the room?

Chapter 6

ESP's greatest magician

6.1 Review

In the last chapter, we saw how magicians can plausibly spot each other using esoteric ESP. Essentially, you simply live your life according to your esoteric beliefs, sprung from the supposed secret to success. Then, your fellow believers can spot your magical status better than nonbelievers.

But there's a catch. Because it takes time to spring magical beliefs from the supposed secret to success, neophytes might not be able to spot more advanced magicians. Neophytes simply are less aware (or not aware) of the beliefs more advanced magicians live by.

Wouldn't it be nice though, if you could spot the grandmasters, though?

6.2 ESP, the coffee shop

I am typing these words while sipping kava at ESP, the coffee shop. ESP, the coffee shop, is the home base for many magicians. I estimate magicians are in the majority here. So, who is ESP's greatest magician? Yes, everyone knows everyone, but, as described earlier, it's difficult to identify the greatest magician in the room.

Well, what if we put it to a simple vote? I don't think that would work. The problem is that the many novices might not vote for the greatest magician.

Instead, I propose we throw a party. Skeptics are invited too! Here's how it works. When you enter the party, the host gives you 100 tokens. You use those tokens to vote for magicians who you trust are greater than you. Say you think Giancarlo and Erda are greater; you divide your tokens between them. While you might not be able to spot grandmasters, it's clear Giancarlo and Erda are at least a little bit more advanced than you.

When giving away your 100 tokens to Giancarlo and Erda, it could be a 50-50 split, or something else. Just vote with your heart. And whenever you receive tokens, just give them away.

This way, tokens flow throughout the party, and at the end of the night, when the gong randomly sounds, you are surprised to discover Josephine holds the most tokens; Josephine is the greatest magician! No one knew except for Abir and Ebenezer, because Josephine had been pretending to be a skeptic the entire time!

There are at least a few problems with this approach though. For example, handing arounds tokens is burdensome, and difficult to do secretly. Also, what if the gong had sounded right after Josephine had handed her tokens to Ebenezer?

Therefore, I propose we use computers to tabulate votes; I propose we use the *PageRank* algorithm.

6.3 PageRank

Twenty three years ago, two Stanford computer scientists formed a company called Google, and they quickly dominated the market for Web search. Their not-so-secret sauce was *PageRank*. According to rumor, PageRank still forms the backbone of Google today.

In the beginning, Stanford patented PageRank, and ended

6.4. EXPERIMENTAL SETUP

up leasing it to Google—in a deal that won $336 million for Stanford. Coincidentally, Rhonda Byrne's book *The Secret* can be purchased today for $10, and she has sold around 35 million copies—yielding an estimated $350 million in sales.

Anyway, here is how PageRank would work to identify ESP's greatest magician. Each person at the party would cast a single paper ballot. On the ballot, you simply list the names of every person who you believe is a greater magician than you. Then you hand your ballot to a trustworthy computer scientist, the scientist enters the data into a computer, runs a quick piece of software, then announces the identity of ESP's greatest magician. Cool, right?

PageRank basically works by simulating a party full of people passing around tokens.

6.4 Experimental setup

Let's simulate a party! Let's begin by making several simplifying assumptions.

(1) We use the PageRank algorithm with $damping = 0.85$, initial PR values of 0.25, and we use 1000 iterations to compute the fixed point.

(2) There are N players, each with a *true* ranking between 0 and 3. There is one player with rank-0, and that player is the greatest magician in the room. However, players do not know each other's rankings. In general, the lower your ranking, the greater of a magician you are. There is one player with rank 0, and 4 players with rank 1, and 8 players with rank 2, and so on. In general, the number of players of rank R equals $\max(1, 4R)$.

(3) A *good* vote is when a player votes for a better-ranked player. A bad vote is when a player votes for a worse-ranked player. Each player casts a random number of votes, such that the following invariant holds: $b = \lfloor g \times \beta \rfloor$, where b is the number of bad votes, g is the number of good votes, and $\beta >= 0$ represents how confused the player is. If β is zero, then the player casts only good votes. If $\beta > 1$ then the player casts

more bad votes than good votes. Etc.

(4) A player's set of best votes may only go to players who are one-rank above. This represents the scenario where magicians cannot identify significantly more advanced magicians, because they don't know the teachings they are living by.

6.5 Experimental results

After conducting several experiments I witnessed that if $\beta < 1$ then PageRank always identified the greatest magician. But once $\beta = 1$, then the accuracy of PageRank goes down. For example, after running 50 randomized experiments, with $\beta = 1$, the greatest magician was only identified 96% of the time. With $\beta = 1.5$, only an 86% success rate.

I think it's interesting that high beta values can still produce accurate results most of the time. Perhaps there's a theorem to be proved regarding PageRank as a method for identifying the greatest magician in the room.

Chapter 7

Rationality

7.1 Review

Thus far, this book has been progressing through a linear narrative. It began with: magicians especially appreciate collaborations with each other, but might face discrimination from skeptics (Chapter 1). Next, magicians can use ESP to keep their magical status secret from skeptics, while revealing their magic to fellow magicians, which can help magicians collaborate with everyone (Chapters 3 – 5). Finally, the last chapter showed how magicians can overcome a limitation of ESP by using the PageRank algorithm (Chapter 6).

This chapter, however, takes a sharp turn. Rather than progressing the linear narrative, this chapter begins introducing preliminary material necessary for understanding *magician's rationality*—a technique magicians can use to guarantee success, in certain situations (Chapter 9).

7.2 Introduction

Personally, I love the concept of *rationality*, and I aspire to be rational—to be logical, reasonable, & wise. If I could pull it off, I would love to take every observation I've ever made, transform these observations into a perfect mind, and then make beautiful, fruitful, fearless decisions—decisions that help me, and help others. For me, rationality is not devoid of magic, rationality is rich with magic.[1]

However, in the field of *game theory*, mathematicians have taken the term "rationality" and they use it in a peculiar way. When mathematicians use the term "rationality," what they usually mean is something along the lines of "perfectly narrow-mindedly selfish."

Thankfully, we can be creative with language. But, perhaps they could have invented a new word, say "miserality," instead of abusing an existing noble word such as rationality?

This chapter, and several more chapters ahead, are all about game theory. Thus I use the terms "rational" and "rationality" often. To highlight my disdain for game theory's definition of "rationality," I was tempted to put quotes around the term "rationality" all throughout this book. However, that's just too many quotes. Instead I just sprinkle quotes here and there, as an occasional reminder of the peculiarity of game theory's definition of rationality.

That concludes my petty rant on the linguistics of game theory. I will stop now, and explain why rationality is an important part of this book.

A few chapters forward, in Chapter 9, we discuss *magician's rationality*. According to magician's rationality, rational magicians are guaranteed to cooperate with each other, to achieve individual and mutual success. However, such success is only guaranteed under certain conditions.

What's really interesting here is that "rational" non-magicians fail to succeed in scenarios that magicians thrive in. Thus, we must first discuss rationality before we can discuss magician's rationality, and that's what this chapter is about.

We begin this chapter with a frightening story that takes place in a dystopic future. This story presents a classic logical dilemma known as the *prisoner's dilemma*, which has disturbed scholars since it was first discovered in 1950.

---

[1] If you're curious about the rationality of holding magical beliefs, just skip ahead to Chapter 13.

7.3 The prisoner's dilemma

You are walking to work one day, when suddenly you are snatched up by the police, for no good reason. Two interrogators take you to a dark room, and one of the interrogators says: "We want you to snitch on your accomplice."

You cry back, "I'm innocent! I don't have an accomplice!"

The one interrogator who is standing says, "We don't care. We want you to snitch anyway."

You pause and think. Then you ask, "Who's my supposed 'accomplice' anyway?"

The interrogator shouts, "That's none of your business!" and he slams the table with his palm, jolting you back.

Another interrogator, who is sitting at the table with you, gently leans forward, and whispers, "We want you to snitch."

"Why should I snitch?"

"We'll offer you a deal."

"What's the deal?"

He says, "It's pretty simple, but pay close attention. If you snitch, and your accomplice doesn't, then you go free."

"And what if I don't snitch?"

"If you *don't* snitch, but your accomplice snitches, then you go to prison for three years."

Yikes. So far, snitching sounds pretty good. But questions remain in your mind.

You ask, "What if we both snitch?"

"Each of you serves two years in prison."

Hmmm.... that doesn't sound too good.

You ask your last question, "What if neither of us snitch?"

"Then you each go to prison for one year."

That's the prisoner's dilemma, summed up in Table 7.1, on the following page. This dilemma has disturbed thinkers for decades. So, what should you do?

Who snitches?	Who goes to prison?
Neither of you snitch	1 year each
You both snitch	2 years each
You snitch, they don't	0 years for you, 3 years for them
They snitch, you don't	3 years for you, 0 years for them

Table 7.1: Choices and consequences in the prisoner's dilemma

7.3. THE PRISONER'S DILEMMA

What might your accomplice do?	What might you do?	Your prison time
They snitch	You snitch	2 years
They snitch	You don't	3 years
They don't	You snitch	0 years
They don't	You don't	1 year

Table 7.2: To snitch, or not to snitch

7.3.1 "Rational" solution

This dilemma was first conceptualized in 1950 by a pair of mathematicians, to help figure out nuclear-warfare strategy for the United States. According to this pair of mathematicians, the "rational" solution is to always snitch. To understand their rationale, let's quickly study Table 7.2, above. This table presents the same possibilities as before, just in a different format now.

Using Table 7.2, observe that your "accomplice" has two choices: to snitch or not to snitch. And now, for the key point, drumroll please... *regardless of which choice your accomplice makes, your best countermove is to snitch!*

OK, to drive this point home more concretely, let's perform the same analysis again, but this time with a microscope upon Table 7.2.

> (1) Suppose your accomplice *snitches*. We are now looking at the *top* two rows of the table. Because your accomplice snitches, you will either serve 2 years or 3 years—no other options are possible. Thus, in this scenario, you should snitch, so you serve 2 years instead of 3.

(2) Suppose your accomplice *does not snitch*. We are now looking at the *bottom* two rows of the table. Thus, you will either serve 0 years, or 1 year. So, you should snitch, so you get away completely free.

Therefore, it seems your best move, the "rational choice" is to snitch, regardless of what move your opponent makes. However, this notion of "rationality" is dysfunctional!

7.3.2 Rationality is dysfunctional

In the previous section we established that "rational" players always snitch. Thus when two "rational" players play against each other, they both snitch, and *they both end up in prison for two years, each!*

That is unfortunate, because if both players were to refrain from snitching, they would both be better off, because then they would only serve *one year each, instead of two years each!*

Thus, the mathematical notion of "rationality" is clearly dysfunctional; which is to say, rational players fail to achieve individual success and mutual success. Because of this dysfunction, the "rational" approach to the prisoner's dilemma has haunted thinkers for decades.

7.4 Conclusion

Wouldn't it be nice if it were "rational" to *not snitch?* Then, "rational" players could achieve individual and mutual success instead of serving unnecessarily long prison terms.

Fortunately, magicians can employ *magician's rationality* to guarantee individual and mutual success, under certain conditions. But before diving into magician's rationality in Chapter 9, we must first cover some more preliminary material in the next chapter (Chapter 8), which presents a philosophy of the interconnection between beliefs, behavior, and logic.

Chapter 8

Deeply believing in success

8.1 Review

In the last chapter I presented game theory's wretched definition of "rationality." Basically, rationality, in this context, means perfect narrow-minded selfishness. But the joke's on game theory, because "rational" players fail to achieve success in games such as the prisoner's dilemma.

In the next chapter, we discuss how magicians can achieve success in games similar to the prisoner's dilemma. But first, we need to discuss some more preliminary material.

8.2 Introduction

The chapter after this one presents *magician's rationality*, a technique of sorts magicians can use to guarantee success in certain situations. Magician's rationality depends upon magicians *deeply* believing in success, and then behaving according to their deep belief in success.

This chapter discusses a bit of philosophy about how beliefs determine behavior, and then describes how a magician can turn a *shallow* belief in success into a *deep* belief in success, and then behave accordingly.

We begin with a mini-essay on how beliefs determine behavior.

8.3 Beliefs & behavior

What does it mean to *deeply* believe something? I contend that if you deeply hold a certain belief, then you will behave according to that belief—you will behave as if your belief is true.

For example, if you were to deeply believe there is a scary spider crawling through your hair, you would certainly do something about it. Yes, it can be that simple. However, I think a more complex example will be helpful.

Suppose I believe that if I visit the city of Las Vegas, then I will become a millionaire. And, furthermore, I believe becoming a millionaire is my most important goal in life. It would therefore be logical for me to visit Las Vegas. Likewise, it would be illogical to refrain from Vegas.

Now, there's a few gotchas with this kind of reasoning. For example, what if my belief is shallow? As in, what if I *believe* Vegas will make me a millionaire, but I don't *deeply believe* Vegas will make me a millionaire? Perhaps I even profess to myself and the world, "Vegas will make me a millionaire!" But, in such a situation, although I might be a perfectly logical person, I might not make the trip to Vegas, because I don't *deeply* believe Vegas would make me a millionaire.

So, how can you really tell if you really deeply believe something? Well, you can figure out what you deeply believe by observing your behaviors. For example, if I don't go to Vegas, then I certainly don't *deeply* believe "Vegas will make me a millionaire" (or I don't deeply believe becoming a millionaire is that important).

Your beliefs are revealed by your actions, not by your thoughts before acting. Interestingly, Douglas Hofstadter, who discovered superrationality,[1] once said something sim-

---

[1] See the Appendix for thorough discussions of superrationality.

ilar: "Your decision is revealed by your actions, not by your words before acting! [1]"

8.4 Logic

There's another gotcha to my reasoning about beliefs and behavior. Specifically, logic is important. The gist is it requires a certain amount of logic to connect your beliefs to your behavior.

Continuing the Vegas-millionaire example, while I might deeply believe Vegas would make me a millionaire, and while I might deeply believe becoming a millionaire is all important—it actually requires logic (a small amount of logic) to connect the dots and realize: I must go to Vegas.

If I am not sufficiently logical, I won't connect the dots, and I might not go to Vegas—thus my behavior might not reflect my beliefs.

While the logic is simple in this case, sometimes the logic might be more complex—perhaps fairly complex, perhaps highly complex. For example, it requires just a little bit of logic to realize you and your fellow magicians will succeed (at least that's what you should believe if you're a magician), and that's what the next section is about.

8.5 The supposed secret to success

Throughout most of this book, we assume magicians *deeply* believe in the supposed secret to success, which is to say magicians deeply believe "if you believe you will succeed, then you will succeed."

Furthermore, we have been implicitly assuming magicians also *at least shallowly* believe that they will succeed. So there's a deep belief in the supposed secret, and perhaps a shallow belief in success.

Here, I am not saying magicians start out *deeply* believing they will succeed. But, at the very least, they at least shallowly believe they will succeed.

So, when you logically combine those two beliefs (a *deep* belief in the supposed secret, and at least a *shallow* belief in your own success)—you acquire a *deep belief that you will succeed*. How does this work, why is this impressive, and why does it matter?

Let's investigate these questions through a hypothetical scenario. Suppose you don't particularly believe in yourself a great deal. Perhaps it's because you believe your past failures point to future failures. Or, maybe your social circle doesn't believe in you, so you don't either.

Regardless, say you come across a book about the supposed secret to success—and it makes sense. Maybe it's this book, or maybe it's another book, but either way you read the book, or you read many books on the supposed secret to success. At first, you don't necessarily believe in magic, but you're at least open to the possibilities. You experiment with the supposed secret in life, and you learn from your successes and failures. And so on.

Through these chains of events, in this scenario, you eventually become a deep believer in the supposed secret to success. You deeply believe in the supposed secret, even though your psychology nags at you, heckling you, telling you that you are more likely to fail than to succeed.

In this scenario, you *shallowly* believe that you will succeed, yet you *deeply* believe in the supposed secret to success. In this scenario, I contend the supposed secret transforms your shallow belief of success into a deep belief of success—supposing you are sufficiently logical. How so? The answer is "modus ponens."

8.6 Modus ponens

Modus ponens is a particular type of logical jump. It is a jump everyone routinely performs, all the time. And, we've been making these jumps for thousands of years. I'm about to explain how it works, and when I do, you may be surprised there is such a fancy term for it.

Suppose I deeply believe, "If it rains, then I will dance in the rain." Then suppose it rains. If I am sufficiently logical, I will go outside and dance in the rain (to be consistent with my beliefs). The bit of logic that connects the *if* part to the *then* part, is known by logicians as *modus ponens*. It's a fancy term, but really it's simple.

With the logic of modus ponens, whenever the *if* part becomes true, then the *then* part becomes true—at least for deeply held beliefs, and if the believer is sufficiently logical.

8.7 Modus ponens and the supposed secret to success

Suppose you deeply believe, "if you believe you will succeed, then you will succeed." Furthermore, suppose you at least shallowly believe you will succeed. According to the logic of modus ponens, you will arrive at the deep belief that you will succeed.

Congratulations, you have just used a deep belief in the supposed secret to success to transform a shallow belief in success, into a deep belief in success. If this truly is a deep belief, then you will behave accordingly.

Perhaps other people deeply believe in you, too. It could be friends, strangers, co-workers, mentors—people who really believe in you and collaborate with you, all behaving under the assumption you will succeed. Perhaps this insight reveals a bit about how the supposed secret works. But it's still magic to me, so who's to say.

Now, my logic here is consistent regardless if the supposed secret is accurate or mistaken. Either way, you end up deeply believing in yourself. And, of course, if the supposed secret is actually true, then you will succeed.

8.8 Conclusion

In this chapter we discussed a philosophy of belief, logic, and behavior. There are several key points: (1) I contend beliefs determine behavior, (2) logic connects beliefs to behavior, and (3) the supposed secret to success transforms *shallow* beliefs in success into *deep* beliefs in success, at least for logical magicians who *deeply* believe in the supposed secret to success.

Putting those three together, when magicians actually, truly, deeply believe in the supposed secret to success (and if they have at least a little bit of a belief in success), then magicians will behave in a manner consistent with success. Now, what does this behavior look like? We dive into this question in the next chapter.

Chapter 9

Magician's rationality

9.1 Review

Let's recap several of the previous chapters. In Chapters 3 – 5, we discussed how magicians can identify each other by witnessing each other's esoteric behaviors.

Then, in Chaper 7 we discussed game theory's peculiar notion of "rationality." We witnessed how "rational" people fail to achieve individual and mutual success in certain scenarios, such as the dystopic prisoner's dilemma. Thus, rationality is dysfunctional.

In the prisoner's dilemma, the police are trying to get you to snitch on an unknown "accomplice." If you are "rational" you will snitch, because it minimizes your prison time, regardless of which choice your "accomplice" makes. Thus, if both players are rational, they both snitch on each other, and they each serve an unnecessarily long prison term. This result is dysfunctional, because if both players behaved "irrationally" they would have fared better—serving only one year in prison, each.

Next in Chapter 8, we discussed a philosophy of belief, logic, and behavior. There are several key points from this philosophy: (1) beliefs determine behavior, (2) logic connects beliefs to behavior, and (3) the supposed secret to success

CHAPTER 9. MAGICIAN'S RATIONALITY

transforms *shallow* beliefs in success into *deep* beliefs in success (at least for logical magicians who *deeply* believe in the supposed secret to success). Thus, logical magicians behave consistently with their deep beliefs in success.

9.2 Introduction

Wouldn't it be nice if it were "rational" to *not snitch?* Then, "rational" players could achieve success instead of serving unnecessarily long prison terms. Fortunately for us, various scholars have discovered alternative concepts of rationality that lead to success.

For example, in 1983, Douglas Hofstadter revealed his discovery of "superrationality," which achieves success under certain conditions. More recently, in 2019, a group of scholars introduced the world to "hyperrationality," which incorporates notions of altruism into rationality.[1]

But, how does the supposed secret to success interplay with rationality? Let's investigate through a dystopic tale from the future.

9.3 Prisoner's dilemma with a twist

It's the prisoner's dilemma once again, but this time there's two new twists: (a) you're a magician and (b) you have the opportunity to meet your "accomplice."

You've been sitting, alone, in your cell for hours, contemplating the deal the interrogators gave you (summed up in Table 9.1).

There's a drip from the ceiling, but no bucket, and mold has been marching towards your dry corner for years. You're trying to figure out a way to use the supposed secret to help

---

[1] See the Appendix for thorough discussions of superrationality and hyperrationality.

9.3. PRISONER'S DILEMMA WITH A TWIST

Who snitches?	Who goes to prison?
Neither of you snitch	1 year each
You both snitch	2 years each
You snitch, they don't	0 years for you, 3 years for them
They snitch, you don't	3 years for you, 0 years for them

Table 9.1: Choices and consequences in the prisoner's dilemma

you succeed, but all routes lead to failure... Suddenly the bars of the cell door swing open with a clank, and two guards toss someone into your cell.

The new inmate tumbles, and lands in the dank corner. He looks up, revealing a confused haze across his eyes.

One of the guards says from behind a plain, red mask, "Talk it over you two. Do your job. Snitch on each other. Got it?"

The guards slam the bars shut, and walk away.

A moment passes. "Are you OK?" you ask your "accomplice."

"I think so. I just need a second," he says. He stands up, and shakes out his arms, breathes in deeply, and says, "OK, I'm ready. Let's talk business."

He tucks his hands inside his arm pits, and starts flapping his arms in a chicken dance, then begins gobbling the Magician's Turkey Wail, which was first established at the Spellmasters International Convention of 2032. The soft gobble gobble melody of bittersweet joy, the pains of illusory defeat, and the triumph of humanity rings through the cell.

Tears well up in your eyes, and you join him in the gobble gobble Magician's Turkey Wail.

Several minutes later, the song ends, you sit down almost completely confident he is a magician. You spend the next several hours chit chatting about each other's adventures, lessons learned, triumphs, psychology, literature, sports, almost everything. Now you're confident.

In the end, you don't snitch on your "accomplice," because you believe magicians will succeed, and to betray a magician would betray your own belief in their success. He doesn't snitch on you, for the same reason—and you each only serve one year in prison.

9.4 Analysis of tale

How do magicians fare in the prisoner's dilemma? Perhaps the supposed secret to success comes to the rescue?

Well, unfortunately, from a strictly skeptical, logical perspective, the supposed secret to success does not guarantee success in the *classic* prisoner's dilemma.

It's just not a scenario where magician's rationality applies. Magician's rationality only works in certain situations. The *classic* prisoner's dilemma is not one of those situations.

The issue is: magician's rationality only works if both magicians both believe each other are magicians. Yet in the classic prisoner's dilemma, your "accomplice" is totally unknown, so magician's rationality cannot apply.

However, if we add a twist to the prisoner's dilemma, then magician's rationality applies. In this twist of the game, we allow prisoners to have conversations with each other, before making their decisions. This way, magicians might be able to identify each other as magicians, using ESP for example.[2]

Suppose two magicians, Alice and Bob, are playing the conversational prisoner's dilemma. They have a conversation. Through their conversation, they confidently identify each other as magicians (say, via ESP). What now? How do Alice and Bob leverage the supposed secret to success to actually achieve success?

---

[2] See Chapters 3 – 5.

9.5 Every logical magician believes every logical magician will succeed

By definition (by my definition), every magician believes in the supposed secret to success, which says, "if you believe you will succeed, then you will succeed." As we saw in Chapter 8, if you deeply believe in the supposed secret, it will transform your shallow belief in success, into a deep belief in success. And, furthermore, if your belief in success is sufficiently deep, it will determine your behavior.

Now for the fun part. Suppose Alice is a logical magician, and Alice deeply believes Bob is a logical magician. Alice will deeply believe Bob will succeed, because Alice deeply believes Bob is using the supposed secret to achieve personal success. Thus... get ready for it... if Alice sufficiently believes Bob will succeed, then this belief will determine her behavior—this belief will ensure Alice does not prevent Bob's success! This logic forms the essence of *magician's rationality*.

9.6 Magician's rationality

Here it is. If Alice and Bob are both logical magicians, and if they both sufficiently believe in the supposed secret to success, and if they both sufficiently believe each other are logical magicians, and if other conditions are met,[3] then their beliefs will ensure they do not prevent each other's success.

If you were to prevent a fellow magician's success, then your behavior would be contradicting your own beliefs—specifically, the belief that your fellow magician will succeed.

Yes: it is logical for magicians to help each other succeed, because they believe each other will succeed. Therefore, whenever mutual success depends upon mutual help, then mutual help is logical, and therefore mutual success is logical. Thus if the magicians are logical, success is logically guaranteed. Magician's rationality is thus based on the premise that beliefs determine behavior; if Alice deeply believes Bob

---
[3] See Chapter 10.

will succeed, then she will behave in manner consistent with Bob's success.

To sum it all up, if Alice and Bob agree two or more years in prison constitutes failure, then neither of them will snitch. Alice will not snitch, because snitching would guarantee Bob serves at least two years in prison, and Alice deeply believes Bob will serve less than two years. Bob thinks similarly to Alice, thus neither of them snitch, and they both achieve success: one year in prison.

There are several interesting caveats to the logical obligation to help—or, rather, the logical obligation to at least not hurt. We discuss these caveats in the next chapter.

9.7 Proof for magician's rationality

See the Appendix for a formal proof of magician's rationality, using fancy symbols such as φ.

Chapter 10

Caveats to magician's rationality

10.1 Review

In the last chapter we saw that it is illogical for magicians to betray each other in a certain situation—namely, the "conversational" prisoner's dilemma. Therefore, logical magicians do not betray each other in that situation, therefore success is guaranteed for logical magicians in that situation. However, there are caveats to magician's rationality. When does magician's rationality apply? When doesn't it apply?

10.2 Introduction

Magician's rationality is a little delicate; it can only guarantee success under certain conditions. Specifically, I have discovered nine conditions which guarantee success. If these nine conditions are met, then magician's rationality will guarantee success.[1] But, before enumerating these nine conditions,

---

[1] Note to reviewer: I am still finalizing my logical proof of magician's rationality (see Appendix). These conditions may change, depending on what I discover through the proof process.

let's first have an informal discussion of a few scenarios where magician's rationality does not guarantee success.

10.3 Zero-sum games

Imagine a game, where there are two players—both magicians. It's your turn and you must decide whether or not to betray your fellow magician.

If you betray your fellow magician, you succeed and they do not. Or, if you refrain from betrayal, then they succeed and you do not. Either way, someone succeeds, and someone does not, therefore mutual success is not possible, and magician's rationality won't help.

In game theory, these types of games are called *zero-sum games*. Magician's rationality doesn't help here, because mutual success is impossible.

Zero-sum games seem to abound in the real world. For example, suppose you and a fellow magician are applying for the same job, and there's only one opening. Should you try your best at the interview? Would that betray your fellow magician?

Well, personally, I think that in the real world (beyond mathematical models of games), we have the opportunity to define our definition of success. So, perhaps it's a little narrow-minded to define "success" as getting one specific, exact job. What does "success" really mean?

10.4 Paternalism

Imagine you are playing a zero-sum game (one winner, one loser) against a fellow magician. Except this time, you believe your fellow magician is a *confused* magician. Specifically, you think your fellow magician has a terrible definition of personal success. In fact, you believe they will fail if they achieve their personal definition of "success." Perhaps they want a cigarette from you, and you don't want them to become addicted.

In this situation, it is actually logical, via magician's rationality, to "betray" your friend. Congratulations, you get to smoke that cigarette they wanted from you. You're really just helping them succeed, as your thoughts go.

This scenario exemplifies *paternalism*—thinking that you know better than someone else, and denying them their definition of "success" is good, because you know better.

10.5 Lack of sufficient logic or knowledge

Suppose you are the parent of a little girl named Kata, who has a science-fair project due next week. Further, suppose both you and Kata believe in the supposed secret to success. Therefore, you believe Kata will succeed at the science fair, and Kata believes too. Therefore, since you believe her success is guaranteed, you don't waste your time helping her. Oops, Kata fails at the science fair. What went wrong?

You were not sufficiently "logical" (or "knowledgeable") to realize how you need to behave to assure Kata's success. If you had been sufficiently logical and knowledgable, you would have realized she needed your help, then you would have helped her.

Here, *knowledge* refers to believing certain facts about the real world. And *logic* refers to the ability to connect dots to draw conclusions from those certain facts. In particular, you need to correctly be able to determine how you should behave to assure success.

10.6 Realism of model

These three scenarios do not disprove the supposed secret to success or magician's rationality, in general, in the real world. The main takeaway, I think, is that you can't always get what you want. If your definition of success is "I always get what I want," then becoming a magician can't guarantee success.

So, what would be a good definition of success? Are there reasonable definitions of success that magic always can truly guarantee, in the real world? For my opinion, you can read between the lines of this book, but, other than that, I prefer to remain silent on this topic.

10.7 Nine conditions to guarantee success

Having discussed three scenarios where magician's rationality does not apply, let's look at the nine conditions that guarantee success amongst a group of magicians.

1. Each magician must at least shallowly believe in their personal success

2. Each magician must sufficiently deeply believe in the supposed secret to success

3. Each magician must sufficiently deeply believe every magician actually is a magician, who sufficiently deeply believes in the supposed secret to success and at least shallowly believes in their personal success

4. Each magician must be sufficiently logical. E.g. each magician must be logical enough to believe that whoever sufficiently believes in the supposed secret to success will succeed

5. Each magician must sufficiently believe every magician is sufficiently logical

6. Each magician must have a personal definition of "success"

7. Each magician agrees with every other magician's personal definition of success

8. Individual success and mutual success are possible

9. The magicians must be sufficiently logical and knowledgeable, to realize how they need to behave in order to achieve success

If these nine conditions are met, then every magician in the group is logically obligated to help everyone achieve success— or, at least they are logically obligated to not thwart anyone's success. The magicians help each other to remain consistent with their beliefs, since we assume beliefs determine behavior.

But, if any of the conditions are not met, then magician's rationality cannot guarantee success. For example, if a magician doesn't believe a fellow magician is a magician, then they won't necessarily believe their fellow will succeed, therefore they are not logically obligated to help. And so on. It's easy to imagine scenarios where magicians won't end up helping each other.

Lastly, it is possible for magicians to succeed *without* meeting these nine conditions. For example, a group of magicians might just be helpful people in general. However, if the conditions happen to be met, then magicians are logically obligated to help each other.

10.8 Conclusion

In closing, I do not believe we need to attempt to persuade magicians to be logical, to live by their beliefs, and to help each other succeed. I believe we are already there. It seems to be intuitive.

I became a magician nine years ago, and since then have witnessed a tremendous amount of magic. I have received so much generosity from my fellow magicians, and I have given it out, too. Helpfulness doesn't need to become our way. It is our way, and I invite you to join us.

Chapter 11

The Zen lemonade garden

11.1 Introduction

As Chapter 9 showed, in certain games, magicians help each other out, to remain consistent with their beliefs. Thus, in certain games, magicians are obligated to help—or, at the very least, magicians are obligated not to prevent the success of their fellow magicians.

But what happens when magicians are *bombarded* with requests for help? How could you possibly help everyone? How could you balance your personal interests with helping your fellow magicians? And, in a "worst case" scenario, what if an army of undercover antagonists were to bombard magicians, trying to undermine magic by wasting everyone's time?

This chapter tells the fictional story of the Zen lemonade garden—a legend about riddles, undercover skeptics, and magicians being bombarded for help. Then, the following chapter, Chapter 12, presents the logical crescendo of this book, which analyzes the logic of this story using computer simulation.

11.2 The legend of the Zen lemonade garden

Professor Digby set his pen down, and glanced up towards the morning sun floating above the calm rippling ocean. He tried squinting, but even then the sun was still to painful for his eyes. So, he stepped up from his stool perched near the stern of the Cheery Mayweather. He began pacing while daydreaming of half-baked theorems, proofs, and lemmas.

Three months prior, Professor Digby had been rejected for a promotion at Oxford—rejected for the position he had been hustling for and pushing for all his life. He lost his composure, to say the least. To more accurately put it, I might say he "freaked out."

So, Prof. Digby cashed out his life savings, and committed himself to a journey to Japan. He planned to hunt down the *Great Zen Professor*, and learn his secrets. His colleagues feared for Prof. Digby. Had he gone mad?

Many of Prof. Digby's colleagues didn't even believe that the Great Zen Professor actually even existed at all. The tales were too tall. For example, according to rumor, the Great Zen Professor only spoke in riddles, yet had managed to publish 500 academic papers the previous year. All this success during a civil war in Japan? Impossible.

But Prof. Digby believed. He had nothing else to believe in, anyway. And if the Great Zen Professor turned out to be a hoax—well, Prof. Digby didn't exactly have a backup plan. He was all in. Maybe he could live a modest life as a fisherman, he daydreamed...

Suddenly, an idea for a lemma appeared in Professor Digby's mind. He squatted down near the bow, trying to hold his notebook in his lap, while he scrawled pen to paper. Maybe the Great Zen Professor will appreciate this proof, he thought. If he really exists, that is. Yes, Professor Digby had his doubts.

11.2. THE LEGEND OF THE ZEN LEMONADE GARDEN

"Land ho!" trumbled a voice from the crow's nest. "Land ho!" Prof. Digby hustled out from the cabin, and there it was. Faint. But in the distance, under high noon sun, there was a tiny silhouette of a mountain on the horizon.

Prof. Digby gripped a beam and leaned in. He felt his stomach turn, then lurch, and he vomited across the deck.

A sailor shouted, "Again!?"

Face down, while still heaving, Prof. Digby gave him the bird. His adventure was about to begin.

Days later.

Haggard, exhausted, and with hay straw stuck here and there across his muddied clothing, Prof. Digby approached the Great Zendo atop Mount Hiei, half a mile above sea level. The doorman guarding the grand entrance surveyed the professor, with cautious eyes.

Prof. Digby thought to himself, *this is it*. Maybe the doorman will turn me away, and I'll become a fisherman. According to the legends, almost no one was admitted into the Great Zendo upon their first try.

But once he was close enough, he noticed the doorman was smiling, and Prof. Digby also noticed he himself was smiling, too. The doorman opened the door, and Prof. Digby stepped through.

Sitting in elegant posture atop a zafu, atop a zabuton, atop polished wooden floors, the Great Zen Professor guided a teacup through the air, aiming to sip while eyeing Prof. Digby, who cautiously approached. The Great Zen Professor gestured for Prof. Digby to sit on a zafu across from him. Four bouncers stood at the center of each of the room's four walls.

Professor Digby sat down, then blurted, "What is the secret to your success? How do you publish so many papers?"

CHAPTER 11. THE ZEN LEMONADE GARDEN

There was a pause, then the Great Zen Professor said,

"I cannot explain Zen to you. You are like a teacup...."

But the Great English professor jumped up, and shouted "But I've come all this way! I risked my life to be here!"

One of the bouncers stepped forward, and whispered quietly, "Leave."

Prof. Digby turned to him, and shouted, "No! I don't need riddles! I need to know the secret to..."

And without a moment seeming to have passed, the bouncer had zolted forward, and flung the English professor out a large, open window, and, after passing through the window, he landed on a pile of soft hay, which only added more to the patchwork of hay across his muddy outfit.

Sitting outside the zendo, hunched over, hands in his face, Prof. Digby bawled. The doorman approached, and sat next to Prof. Digby, then said, "I know this is a tough break, Englishman. I'm so sorry it didn't work out as expected. But, I've got good news! The Great Zen Professor told me he will see you again, and explain his secret. All you need to do is earn a *Zen number of two*." The doorman smiled.

Prof. Digby jolted up. "Really? Wait, what's a Zen number? And, how do I earn a Zen number of two?"

"It's simple, Englishman. Co-publish a paper with someone who has a Zen number of one."

"Who has a Zen number of one?"

"Everyone who has co-published a paper with the Great Zen Profssor!"

"So all I have to do, is find someone who's collaborated with the Great Zen Professor, collaborate with them, then the Great Zen Professor will see me again?"

"Pretty much."

"Thank you, thank you, thank you!" and he was off.

11.2. THE LEGEND OF THE ZEN LEMONADE GARDEN

The doorman sat and watched Prof. Digby run towards the trail down the mountain, thinking about how Prof. Digby hadn't a clue about the journey before him. The doorman wished to himself journeys didn't have to be so arduous. But this seems to be the only way. He walked back to his post, whistling.

Prof. Digby eventually arrived at Ryukoku University, near the base of the mountain, and quickly found a professor with a *Zen number of one*. He requested collaboration with the Ryukoku professor. The Ryukoku professor replied:

> "If you collaborate with one of my students, you will earn a *Zen number of three*, then I will collaborate with you, then you will have a *Zen number of two*, then you can collaborate with the Great Zen Professor."

Prof. Digby found a grad student with a *Zen number of two*, who said:

> "If you collaborate with one of my students, you will earn a *Zen number of four*, then I will collaborate with you, then you will have a *Zen number of three*...."

But Prof. Digby interrupted, saying "Where can I find one of your students?"

The grad student smiled, and said, "There's a lemon farmer named Aika on the outskirts of town...." and then he gave directions.

Chapter 11. The Zen Lemonade Garden

As Prof. Digby rounded the final bend of the trail, he finally caught sight of the lemon farm, and to his dismay: there seemed to be at least twenty people standing in line outside the farmhouse. He raced forward.

"Are you here for a Zen number?" Prof Digby asked the last person in line.

The man waved his hand, pointing over everyone in line, and said, "We're all here to earn to earn our first Zen number. Except for the impostors. We, I mean *they*, the impostors, they are just trying to waste everyone's time. I'm not an impostor, trust me."

This is crazy, Prof. Digby thought.

Two hours later, it was finally Prof. Digby's turn. He entered the house. The farmer sat in a wooden chair, in her living room, with scattered papers everywhere, half-eaten bread rolls on the floor, and the smell of must wafting around.

He asked her to collaborate. She said, "Wait." She shuffled around her mess, looking for something. About 30 seconds later, she shouted, "Yes!" picked up a paper, which looked woodblock stamped, containing English words. She handed it to him. He speed read the following:

11.2. THE LEGEND OF THE ZEN LEMONADE GARDEN

英語

Welcome student!

Would you like a Zen number? Let's collaborate!

I will do the least amount of work, and I will receive the honor of *last* author on our published paper.

You will do the most amount of work, and receive the honor of *first* author on our paper.

I can remove my name from the paper, up to the last minute, right before printing.

The topic of our paper will be *the secret to success*.

His heart leaped. He would finally learn the great secret of the Great Zen Professor's success! But, from a farmer?

Prof. Digby asked, "So you're my teacher right? Can you tell me—*what is the secret?*" He glanced around for bouncers.

She said, "Sorry I don't have time to fill you in—there's a bunch of impostors I need to deal with right now. Go on, move it along, go do it."

"Oh yeah, the impostors..."

"Are you an impostor? Are you wasting my time? Doesn't really matter, but in case you're genuine, I'll clue you in. Most of the people in line are trying to undermine Zen by wasting everyone's time. Prove to me you're not an impostor by writing that paper. The impostors never complete their homework. I think they're afraid it would indoctrinate them."

"How can I write the paper if you won't tell me the secret!?"

"Go do it. I believe in you! You should, too. Oh yeah, your paper must be a *koan*, and a poem. Both at the same time."

"But I'm a mathemetican..."

"Leave," she said.

His face boiled, and he shouted, "Where's the bouncer!?"

And then she flung him. He arced through the air, through the doorway, past an impostor next in line, and landed in a soft pile of hay.

Over the next week, everyday, Prof. Digby showed up at the farm, waited in line, received about 5 minutes worth of koans from his teacher, then she sent him on his way.

And every evening, he sat at home, puzzling over the riddles she had given him, trying to piece everything together. As the days passed, the riddles started adding up, making more and more sense. During his seventh evening of self study, a light bulb went off in his head! It was a theorem to describe the purpose of the Zen number system. He scrawled it out over ten pages in his notebook, his hands quivering.

This is it, he thought as he scrawled.

The next day, when it was his turn to enter, he rushed in and shouted "I've got it! I've got it ! I've figured it out! It's recursive and logarithmic! $w = 1 + \log_2(1/t)$ and ..."

But all she heard was gibberish. She was a poet, not a mathematician. But she could tell from his excitement, he had figured it out, so she interrupted him, saying, "Great! Turn it into a poem!" Then, she handed him a lemon, and said, "Would you mind making us a pair of lemonades?"

"That's it!" he shouted!, and ran off, lemons in hand. Then he thought better of it, returned, and made a pair lemonades, which they sipped together, while he played with words in his mind, and she laughed and he smiled.

The next day he returned with a poem, neatly written on a fresh piece of paper. It read:

11.2. THE LEGEND OF THE ZEN LEMONADE GARDEN

If you're busy, and someone hands you a lemon, hand it back, and ask for lemonade.

If you're not busy, make lemonade, and hand it back.

Teach, don't leech.

Believe in yourself and others.

She laughed, shook his hand, and said, "You did it! Let's publish." She taught him how to carve woodblock prints while they minted the poem together into a slice of maple trunk. He had earned his first Zen number—a *Zen number of four*.

Prof. Digby had learned an aspect of the secret to success, the secret of the Great Zen Professor, the secret to publishing countless papers. In his diary that night, Prof. Digby wrote:

> Basically, the Great Zen Professor uses koans (*lemons*) to teach as many students as possible, as quickly as possible, with as little effort as possible. Such artfulness is necessary because an army of impostors seeks to undermine Zen.
>
> Lemons educate genuine seekers (who make *lemonade* from the lemons), and lemons also cast away the phonies (who are never willing to put in the time to make lemonade, because they are afraid lemonade would indoctrinate them into Zen).
>
> The Great Zen Professor delegates a tremendous amount of his teaching load to his students. Thus, each of his student-teachers faces a dilemma: *to teach or to leach*. Sure, students could advance more quickly by refraining from teaching—because then they have more time to solve their koans.

But, if everyone refrains from teaching, then there's no one to teach but the Great Zen Professor, because the whole hierarchy of student-teachers would crumble.

How do student-teachers overcome this dilemma? The solution is for entry-level students to learn about the teach-or-leech dilemma, and then overcome the dilemma by believing in themselves and others.[1] If every student-teacher believes in the success of every student-teacher, then leeching is illogical, because it betrays your belief in everyone's success.

QED.

Prof Digby set his pen down, and pondered his life. There's so much more to learn. There's so much more to *unlearn*.

Prof. Digby became a student of the grad student, and they spent much time together. During the same period, Prof. Digby began teaching—dividing his time in half between making lemonade, and serving lemons to the large numbers of students and impostors that showed up at his cottage every day. He gave everyone little slices of his time, and, as it turned out, some of his students solved Prof. Digby's koan, proving themselves genuine seekers.

Once he published a paper with his grad-student teacher (earning a *Zen number of three*), he began working with the Ryukoku University professor. Somehow, through the magic of the Zen lemonade garden, Prof. Digby was able to teach full time, study full time, and advance quickly, and feel relaxed all along. Pretty soon he published a poem with the Ryukoku professor, and Prof. Digby had finally received his *Zen number of two*.

He was finally ready to meet the Great Zen Professor once again.

---

[1] Magician's rationality, see Chapter 9.

11.2. THE LEGEND OF THE ZEN LEMONADE GARDEN

Professor Digby took a leisurely stroll up the mountain. He had just washed his suit the day prior, and the sun shimmered brightly across every seam. As he rounded the final bend, Prof. Digby witnessed the Great Zendo beaming reflections of the morning sun. The beams must have reached his heart, for a tear welled up...

"Englishman!" shouted the doorman.

"Greetings good sir. I have my papers," replied Prof. Digby.

"Oh yes, oh yes, I see. A Zen number of two, that's great. Oh a binary logarithm, neat. Yes, yes, come on in," and the doorman opened the grand entrance of the Great Zendo.

The Great Zen professor sat upon a zafu, upon a zabuton, upon the polished wooden floor, surrounded on all sides by students sitting on zafus, upon zabutons, upon the polished wooden floor. They turned to see Professor Digby step into the room.

One student waved, and Professor Digby waved back. Then, the Great Zen Professor addressed Professor Digby, saying "We're brainstorming on papers right now, if you'd like to join," and the Great Zen Professor pointed to a vacant zafu.

Professor Digby sat down, quietly and soaked in the warm feeling of optimized creative expression.

A few days later Professor Digby sat in his cottage, carving a theorem into a maple woodblock. It was a formal mathematical theorem which was so tedious to carve, it was silly. A drop of sweat fell off his nose and landed on woodblock, and Digby laughed.

He was about to earn his Zen number of one, which meant his adventure was almost over. He had learned the secret of the Great Zen Professor, he had written a poem about it, and published it. He had taught the secret to students and impostors (who didn't even bother trying to understand). He had been flung by bouncers, and so on his adventure went, obstacles, victories, and the rest. But now it was ending.

As he carved his woodblock, he pondered his next move. Maybe I'll go back to Oxford, and see if they'll hire me to teach Japanese. Maybe I'll stay here and learn Zen. Or maybe I'll become a fisherman, back home or perhaps here.

And when he thought of staying in Japan, he realized he couldn't stay. Sure, he felt very welcomed, and it was comfortable now. And, he enjoyed working with the Great Zen Professor and his students, but...

What was it that caused Digby to set his woodblock down, and dream of home again? Maybe it's because in his daydream, he held a pile of papers that his colleagues back home didn't even believe existed—proof that the legend of the Great Zen Professor is true! Many have speculated that Digby dreamt of vindication, and I see the truth in that theory myself. However, I hold a theory that for Digby's whole life, he had held an unfulfilled thirst for adventure his whole life, and he never realized it, because he never knew anything different. But now, having triumphed in adventure, the desire washed away. Perhaps I'm projecting my own psychology upon this fictional character. Authors are known to do that, and I apologize.

Anyway, Digby placed his woodblock down, and stepped out of his cottage. "If I'm going to set sail for home, I'll need to make some money," he thought as he began his walk to the wharf to become a fisherman.

Chapter 12

Lemons & lemonade

12.1 Review

The previous chapter tells the story of Professor Digby's quest to learn the secret of the Great Zen Professor. How did the Great Zen Professor publish so many papers, so quickly, while being bombarded with requests for help (from friends and foes alike), while only speaking in riddles? This story can perhaps be summed up by a riddle—a poem penned by Prof. Digby upon learning the secret of the Great Zen Professor:

If you're busy, and someone hands you a lemon, hand it back, and ask for lemonade.

If you're not busy, make lemonade, and hand it back.

Teach, don't leech.

Believe in yourself and others.

12.2 Introduction

As Chapter 9 showed, in certain games, magicians help each other out, to remain consistent with their beliefs. Thus, in certain games, magicians are obligated to help—or, at the very least, magicians are obligated to refrain from thwarting the success of their fellow magicians.

But what happens when magicians are *bombarded* with requests for help? How could you possibly help everyone? How could you balance your personal interests with helping your fellow magicians? And, in a "worst case" scenario, what if an army of undercover skeptics were to bombard magicians, trying to undermine magic by wasting everyone's time?

In this chapter, the logical crescendo of this book, we explore these questions by modeling and simulating *koans*, Zen riddles—while tying in the story of the Zen lemonade garden, from the previous chapter.

We begin with *koans*.

12.3 Koans

According to legend, Zen teachers are sometimes bombarded by requests from both genuine spiritual seekers and antagonists, at the same time. The latter are confused people who seek to undermine Zen by wasting everyone's time with collaborations that go nowhere.

Zen teachers can use "koans" to gracefully handle such overloads. The gist of the koan trick is to spend just a little bit of time with each student who shows up at your doorstep. During this time, you communicate a personalized Zen riddle, a *koan*, to the student (who might really be an antagonist), then send them on their way.

The genuine students will eventually solve their koans, and come back again, asking for more advanced teachings, after having learned a lesson through independent self study.

What about the antagonists then? Do they do their homework? I.e. do they solve the koans assigned to them? In this legend (in this model), the answer is no, the antagonists never solve koans for two reasons. First, these impostors would need to spend their own time to solve koans, and they would rather just waste other people's time, than their own. Second, they secretly fear that if they were to solve koans, they would become indoctrinated into Zen.

For koans to work, the riddles must, at the very least, (1) be easy to generate, (2) easy to verify the solution, and (3) should enlighten the minds of the riddle solvers. Consider an example.

12.4 Koan #1

Consider the following koan:

If you're busy, and someone hands you a lemon, hand it back, and ask for lemonade.

If you're not busy, make lemonade, and hand it back.

What does this koan mean? And how long would it take you to figure it out on your own? Let's suppose you haven't read the previous chapter. Then how long would it take you to understand this koan? What if this book was empty, except for this koan? How long then? How could you go about figuring it out?

Allow me to explain the koan. It's a koan about koans.

If you're busy, and someone hands you a lemon, hand it back, and ask for lemonade.

I.e., if someone asks for your help (such a request is a lemon), and if you're busy, then quickly give them a hint in the form of a koan (another lemon), and ask them to come back with proof they've figured it out (lemonade).

Now for the second half of this koan:

> *If you're not busy, make lemonade, and hand it back.*

I.e., if you're not busy, and if someone asks for help (a lemon), then spend time with them, helping as much as you can (making lemonade).

Another way to interpret this half of the poem is: if you're not busy, and no one is asking for your help, then take the koan previously given to you by your teacher (a lemon), and study it, turning it into lemonade, then offer it to your teacher.

12.5 The Zen lemonade garden

The previous chapter tells the story of the Zen lemonade garden. Essentially, a great Zen teacher uses koans to enlighten the minds of genuine students, and to cast away the phonies. In particular, the Zen teacher sets up a hierarchy of helpers, to keep the phonies on the periphery while creating comfortable room for genuine students on the inside. It works using the "Zen number system."[1]

The Zen number system assigns numbers to students and teachers. The great Zen teacher sits at the top of the hierarchy, with a Zen number of 0. Anyone who collaborates with them earns a Zen number of 1. Anyone who collaborates with one of those people earns a Zen number of 2. And so on.

Furthermore, you may only collaborate with the least qualified teacher available. So suppose you earn the Zen number 5 by collaborating with someone with a Zen number of 4. Now, you are qualified to collaborate with someone with a Zen number of 3, to earn a Zen number of 4.

From this system of Zen numbers, a dilemma arises. Why should you bother teaching other students, when you could be spending time advancing your own interest by solving

---

[1] The Zen number system is inspired by the Erdős number system.

koans from your teachers? The following koan addresses this dilemma.

12.6 Koan #2

Consider the following koan:

Teach, don't leech.

Believe in yourself and others.

What does this koan mean? I think this one's pretty straightforward if you are familiar with magician's rationality (see Chapter 9).

The dilemma is to *teach* (i.e., generously offering your time to others, at the expense of your own personal advancement), or to *leech* (to take advantage of the generosity of others, without being generous yourself, by studying while refraining from teaching).

12.6.1 Prisoner's dilemma

This teach-or-leech dilemma is similar to the prisoner's dilemma from Chapter 9. Recall, in this prisoner's dilemma, each player must choose to snitch or refrain from snitching. If both players snitch they spend an unnecessarily long time in prison, but if they can find a way to cooperate (e.g. via magician's rationality), then everyone gets a much better outcome.

12.6.2 Teach-or-leech dilemma

Similarly, in the teach-or-leech dilemma, players must choose to leech or teach. If everyone leeches, then no one teaches, and everyone takes an unnecessarily long time to advance. But leeching is tempting, because if you're the only leech, then you can speed along, while taking advantage of everyone else's generosity. Everyone else is a victim of such leeching, because their teaching workloads increase, since the leecher isn't teaching.

On the other hand, if everyone teaches as much as requested (e.g. via magician's rationality), then everyone advances learns quickly—which I prove via simulation, soon in this chapter.

12.6.3 Solution

So, how do we solve the teach-or-leech dilemma? Magician's rationality, of course! To be admitted into a Zen lemonade garden, i.e. to earn your first Zen number, you must prove to a guardian of the garden that you believe in the supposed secret to success. This way, everyone in the garden is provably a magician, and the student-teachers can take advantage of magician's rationality.

Thus, your first koan, to be admitted into the garden, is to learn to believe in the supposed secret to success.

12.7 Modeling the Zen lemonade garden

I believe the Zen lemonade garden (ZLG) is more than a mere fable; it is useful in the real world. For instance, suppose someone asks you to teach them the secret to your success, but you're already busy being helpful. If you can only spare a second, just point them to the appropriate section at the local public library.

But how realistic is the legend? Do Zen teachers really employ ZLG in the real world? Should they? I would like to begin the effort towards answering these questions by defining a mathematical model of ZLG. Ideally, our model should be simple, precise, and realistic. However, modeling this legend is difficult. For example, I have not scientifically collected data to answer questions such as: how much time does it take for a Zen techer to generate a good koan, supposing they just met the student an hour ago? And, how much time would such a koan take to solve, on average?

Without the aid of real-world data, I searched for a simple and precise model that might be insightful anyway. I think it's pretty good. Let's begin our discussion of my model with koans—specifically receiving, transmitting, and studying koans.

12.8 Receiving, transmitting, and studying koans

My model represents a kind of simplified best-case scenario. The whole model is based on the premise that you divide your workday between three tasks: (1) receiving a koan (if needed), (2) transmitting koans (if needed), and (3) studying your current koan.

So, how much time does it take to receive and transmit koans? And how much time does it take to solve koans? We use the following formula, where w is the number of days required to solve a koan, and t is the number of days needed to transmit that koan (where $0 < t \leq 1$).

$$w = 1 + \log_2(1/t)$$

In the appendix, I present a detailed look at why this formula is worthy. But, for now now, I just want to point out two characteristics of this formula. First, if you dedicate more time to transmit a koan, it takes less time to solve. Second, there are diminishing returns. For example, if you spend a *fourth* of a day transmitting a koan, it will take *three* days to solve. But suppose you spend twice as much effort, spending *half* your day transmitting a koan, then it will take *two* days to solve, which is less. However, even though you have doubled your effort in transmitting a koan, you have *not* halved the effort in solving that koan.

12.9 Finding a teacher

Suppose you need a fresh koan. Perhaps you have just arrived in the garden, or perhaps you have just solved your current koan and need a new one. In the model, you may only receive a koan from the least qualified teacher available. For example, suppose you have solved 5 koans, and there are two practitioners in the garden who have solved 6. You will need to receive your next koan from one of those two practitioners.

Restricting students to least qualified teachers is key to defending against antagonists. Essentially, more advanced practitioners are shielded from any neophyte shenanigans.

12.10 Antagonists

In our model, antagonists never actually solve any koans—they just try to waste everyone's time by requesting them. Thus, antagonistic shenanigans only bother the *threshold guardians*—the most novice teachers. Perhaps the antagonists are afraid that if they actually studied and solved koans they would become indoctrinated into Zen.

12.11 Balancing teaching and studying

In our model, we assume practitioners are self interested, which is to say they wish to study Zen for their personal benefit, by solving as many koans as they can, as quickly as they can. Yes, Zen teaches practitioners to help others for the sake of others, but we model a scenario where practitioners do not believe in such teachings.

So why would a practitioner ever teach, when they could be spending their time solving koans? To begin addressing this question, we need to first discuss the formula for balancing teaching and studying.

At the beginning of each day, each practitioner decides how they will allocate their time. Specifically, they decide how much time to set aside for personal studies. For any remainder, students make themselves available as a teacher (supposing they are eligible to teach, and there are eligible students wishing to receive a new koan).

If one or more students show up at a teacher's doorstep on a given day, and the teacher has agreed to teach, then the teacher divides their teaching time equally amongst the students; the teacher gives each student the largest koan possible, such that the teacher doesn't teach any more than they want to.

12.12 The teach-or-leech dilemma

Practitioners face a dilemma: to teach or to leech. Leeching is alluring, because then you can max out personal interests, instead of spending your time helping others. But, if everyone leeches, then no one teaches except the Zen master, and you will progress unnecessarily slowly.

This dilemma is similar to game theory's prisoner's dilemma (see Chapter 16). To wit, if you teach, and everyone else leeches, they benefit and you do not. But if everyone teaches, everyone benefits. Lastly, if you were the only leech, you could progress the fastest.

How do we solve this dilemma? Magician's rationality, of course.

12.13 Magician's rationality

Suppose a Zen lemonade garden is a school of magic. Zen is magical after all. In other words, suppose every teacher believes in the supposed secret to success; suppose every teacher holds as an axiom "if you believe you will succeed, then you will succeed." Then, in this case, we do not need to persuade teachers to cooperate. Rather, a shared belief in the secret to success logically implies *magician's rationality*—helping each other out, and the practitioners will succeed (see Chapter 9).

12.14 Conclusion

I conducted several experiments and learned that my model of the Zen lemonade garden works well, even when under heavy attack by antagonists. See the appendix for details.

I will reiterate here though that the success of Zen lemonade garden depends upon cooperation. Essentially, students may be tempted to spend more time studying their own koans than spending their time helping out their fellow students. But, if everyone behaves selfishly in this way, then the Zen master will be the only teacher, and everyone will progress unnecessarily slowly. For the garden to perform well, everyone needs to give at least a little, and, as my experiments show, if everyone gives a lot, then everyone receives a lot.

This is where magician's rationality comes in handy. Recall from Chapter 9: it is logical for magicians to help each other in certain situations. Thus, in the Zen lemonade garden, every student's first lesson includes a teaching about magician's rationality.

However, my definition of magician's rationality only applies to certain games, and the Zen lemonade garden is not technically one of those games.[2] Perhaps, there exists a reasonable mathematical definition of magician's rationality for the Zen lemonade garden, and similar games. I don't think it really matters though, for the purpose of this book, because what I really care about is achieving magician's rationality in real-world scenarios.

In the real world, outside of mathematical models, what can you do but do your part to manifest your beliefs into reality—simply by behaving according to your beliefs, by helping your fellow magicians, which also helps you too.

---

[2]TODO: revisit

Chapter 13

Thoughts for skeptics

13.1 Review

Over the previous 12 chapters, we studied the supposed secret to success via magic, logic, and simulation. Our journey walked us through the tension of magician-skeptic collaboration, covert communication via ESP, identifying the greatest magician in the room, guaranteeing success through magician's rationality, and, finally, coping with being bombarded by requests from your fellow magicians.

This chapter, and the following two final chapters, step back from an incremental narrative about magic and logic. This chapter makes the logical case for why it can be reasonable (from a skeptical perspective) to adopt unproven magical beliefs. Then, in the next chapter, I present an analogue for magicians: why it can be reasonable (from a magical perspective) to adopt skeptical perspectives. Finally, the book concludes with Chapter 15, a short story about a little girl haunted by uncountable wishes.

13.2 Introduction

Many skeptics object to unproven magical beliefs. For, according to certain versions of skeptical philosophy, it is unreasonable to believe in unproven beliefs, particularly magical beliefs for which there is no sound explanation of causality. In this chapter, I presume you are one of these skeptics.

My driving question for this chapter is: what if there exist truths that cannot be proven via a strictly skeptical mindset? As in, what if a particular magical belief were true, but it cannot be proven while holding on to strict skepticism? You may think I'm copping out right now. You might be thinking: "Oh isn't that just convenient. The magician can't prove his claim. Surprise, surprise."

I am not copping out. First, as I will soon explain, there certainly exist *mathematical truths* that cannot be proven, period. Would it be so strange if there existed *non*-mathematical truths that cannot be proven via skepticism?

Second, I daresay I can prove to you, prove to your satisfaction, that it can be reasonable to adopt unproven magical beliefs—beliefs that are unfounded in sufficient concrete, verifiable evidence.

I introduce my proof via a famous story from the history of mathematics, the story of Gödel's proof, which begins with the story of *ignorabimus*.

13.3 Ignorabimus

The late 19th century and the early 20th century saw the development of a new backbone for mathematics—a foundational model of math itself, which has since become canonical. During this time period, a philosophical debate arose. On one hand were prophets who believed that their efforts to model math would reveal the truth of *ignorabimus*—the truth that certain mathematical beliefs are *true, yet unprovable*. Essentially, they claimed there exists unknowable mathematical knowledge.

On the other hand, a bunch of other philosophers disagreed. For they believed that there must exist a correct proof for every true mathematical belief. In other words, they claimed that every true belief could be proven.

So, why did the prophets of the ignorabimus camp believe that there are unprovable truths? Let's consider Goldbach's conjecture, first articulated in 1742.

Conjecture 2. *Every even whole number greater than 2 is the sum of two prime numbers.*

To this day, hundreds of years later, no one has proven, nor disproven this conjecture. Many have tried; no one has succeeded. Perhaps, in the spirit of ignorabimus, it is impossible to prove this conjecture using formal mathematics?

13.4 Hilbert and Gödel

On with the story. David Hilbert was a luminary amongst mathematicians. He received his doctorate in in 1885, and spent the next 45 years of his career producing significant contributions to mathematics, during the early efforts to model math iteself. Hilbert was at the forefront of the effort, leading the search for a sort of "theory of everything"—the search for proof that *every true mathematical belief is provable*. In other words, Hilbert led the effort to prove the ignorabimus camp was wrong.

He retired in 1930, and gave his retirement address at an academic conference on the theory of knowledge, which happened to take place in his hometown of Königsberg.

In his remarks, he bid adieu, professionally restating, one last time, for everyone, in public, his belief that every truth is provable:

> For us there is no a ignorabimus, and in my opinion even none whatever in natural science. In place of the foolish ignorabimus let stand our slogan:
>
> We must know,
> We will know.[1]

---
[1] Translated from German by James T. Smith.

I am unfamiliar with Kurt Gödel's sense of humor, who was probably in attendance, but if he were there, perhaps he was chuckling. For, at the age of 24, having just completed his Ph.D. on the topic of ignorabimus, Gödel had proof in his pocket that Hilbert was wrong—mathematical proof that the ignorabimus camp was right! *It truly is true that there exist mathematical truths that cannot be proven!* Not too long after Hilbert's speech, at a roundtable discussion at the conference, Gödel publicly announced his proof.

Today, Gödel's proof is widely accepted as canon.

13.5 Belief as a form of inquiry

Yes, I am poking fun a little bit at Hilbert, however I consider his quest for a theory of everything to be noble. In mathematics, or science, and perhaps elsewhere, belief in unproven claims can catalyze discovery.

To be concrete, Hilbert adopted the belief that his theory of everything was true, then set out to discover proof of its truth. This is one way mathematicians discover proofs. They play around with math, then notice a pattern, then try to prove some sort statement about the pattern. Proving propositions is hard work, and when there's an ocean of possibilities, mathematicians tend to direct their efforts towards proving things they think are true. For if you don't believe your proposition is true, why waste time trying to prove it?

If Hilbert hadn't believed in this theory of everything, he probably would have never even bothered searching for such an ambitious proof. Yes, Hilbert turned out to be mistaken, but here belief acts as propulsion.

13.6 The will to believe

Rewinding in time in our story, all the way back to 1896, a philosopher named William James gave a lecture titled "The Will to Believe." In his lecture, he claimed that certain truths can only be witnessed by believing in those claims *before* being

13.6. THE WILL TO BELIEVE

able to witness the evidence. As in, *you've got to believe it to see it,* as opposed to you've got to see it to believe it. Philosophers have been arguing about his thesis since. This thesis, this conjecture is an unproven magical belief.

Conjecture 3. *For certain truths, you've got to believe it to see it.*

We have just seen that ignorabimus provably applies to mathematics (i.e. Hilbert was wrong), but perhaps a sort of ignorabimus applies to knowledge in general. Specifically, in the spirit of James, perhaps there are truths that must be believed before they can be witnessed. What if there exist truths that can only be proven to oneself after adopting a belief in the unproven truth? Even strict skeptics must admit the will-to-believe thesis is plausible—for, if we are to be strictly skeptical, we lack certifiable proof that the thesis is wrong (even after a century of philosophical inquiry and debate). Let's go further. Let's dive into my theorem and proof for this chapter.

Theorem 2. *It can be reasonable to adopt unproven magical beliefs*

Proof. Can it really be reasonable to adopt an unproven magical belief? Let's consider the two cases concerning the factuality of the will-to-believe thesis (Conjecture 3). Broadly speaking, either (1) Conjecture 3 is wrong, or (2) Conjecture 3 is correct.

Let's begin by considering the second case; i.e. let's tentatively assume that the will-to-believe thesis is correct, just to see where it might lead us, logically. Specifically, if we assume the thesis is true, then *it is possible that the thesis applies to thesis.*

In other words, it might be possible that the only way to witness the truth of the will-to-believe thesis, would be to choose to believe the will-to-believe thesis first, and then see where that belief takes you. Perhaps that belief would enable you to witness the truth of the will-to-believe thesis. Logically, this is a possibility.

Let's go back and consider the first case, i.e. let's consider the case that the will to believe thesis is wrong. Then, if you were to choose to believe the thesis, you would not discover the truth of the thesis.

Either way, one reasonable approach to investigating the will-to-believe thesis is to choose to believe it, and then experiment with it by adopting various new unproven magical beliefs, and seeing what happens. □

13.7 Bibliograhpy

[1] Mathematics Genealogy Project: David Hilbert, https://mathgenealogy.org/id.php?id=7298

[2] "David Hilbert," Physics Today, 2018, https://physicstoday.scitation.org/do/10.1063/pt.6.6.20180123a/full/

[3] Stanford Encyclopedia of Philosophy: Hilbert's Program, https://plato.stanford.edu/entries/hilbert-program/

[4] James T. Smith, "David Hilbert's Radio Address," 2014, http://math.sfsu.edu/smith/Documents/HilbertRadio/HilbertRadio.pdf

[5] Stanford Encyclopedia of Philosophy: Kurt Gödel, https://plato.stanford.edu/entries/goedel/

[6] Stanford Encyclopedia of Philosophy: Notes to Kurt Gödel, https://plato.stanford.edu/entries/goedel/notes.html

[7] Institute for Advanced Study, "Kurt Gödel: Life, Work, and Legacy," https://www.ias.edu/kurt-g%C3%B6del-life-work-and-legacy

[8] William James, "The Will to Believe," http://krypton.mnsu.edu/~jp6372me/THE%20WILL%20TO%20BELIEVE%20.pdf

Chapter 14

Thoughts for magicians

14.1 Review

In the preface, I told you a little bit about how this book came about. To briefly recap, I had a career as a professional skeptic. Then, nine years ago, I faced a psychological crisis, and responded by embracing belief. Over the next of nine years I slowly reacquainted myself with skepticism, culminating in a synergy of belief and skepticism. This book represents a fruition of my journey.

But, there's more to the story. In particular, there is the origin story of the Zen lemonade garden, which I believe might offer valuable insights for magicians.

14.2 The origins of the Zen lemonade garden, Part 1

Recall from Chapter 12, the Zen lemonade garden (ZLG) is fundamentally a method for coping with overwhelming requests for help—particularly when some people may be trying to waste your time.

The inspiration behind ZLG began at MIT, over a decade ago. I worked there as a computer scientist, specializing in computer security. There, I became acquainted with a particular genre of cyber attack: *high-density attacks*. The way these attacks work is you sneak a tiny little request into a victim's computer. Then, your tiny little request wastes the victim's computer time. Such attacks can be devastating.

Disturbed, I set out to discover a defense. In 2010, I joined Harvard as a Ph.D. student, and, through collaboration, came up with a defense I named *bouncer*.

But, the bouncer was only a partial solution, for it only works under certain conditions. But, rather than continuing to pursue my research at Harvard, I just wrote a math proof, received a master's degree, and dropped out with the hope of becoming a fast-money millionaire in Silicon Valley.

I joined Twitter as an early employee, and consulted to DARPA on the side—the United States agency that invented the Internet, stealth airplanes, and GPS. For DARPA, I picked up where my Harvard research left off. With a collaborator, I developed the *doorman*. Putting the bouncer and doorman together produced *Beer Garden*.

Shortly thereafter, I experienced my crisis (see Preface), and left my career. Grappling for a notion of truth, I replaced skepticism with a belief in Buddhism and magic.

Six months later, while sitting on a meditation cushion, my mind popped for the first time. I witnessed a flash of a vision of world peace, and my epic role in it.

Suddenly 100% sure I would succeed at rapidly ushering in world peace, peacefully, I set out to fulfill my duty, and shortly woke up in a psych ward. It's been a long back-and-forth journey, but today I am aware I have schizoaffective disorder and, these days, I believe I manage my delusions reasonably well.

I tell you this now, because I know what it's like to profoundly believe in my ability to succeed at everything. For instance, suppose hypothetically you believe you will succeed at everything. Why not be ambitious? Why not set your sights on world peace? OK, now that you are 100% sure you will succeed at ushering in world peace, many questions arise. Among them: (1) who else realizes you are about to succeed at world peace? (2) Do they want to help or hinder you?

For whatever reason, I presumed there exists at least one powerful, competent organization interested in global affairs. Since they are competent, as I thought, they of course know the epic role I play, they can observe everything I do, and they want to help me, as long as I play by their rules. They sent me secret messages through radio, TV, films, YouTube, friends, family, strangers. They subtly redirected me to help me fulfill our shared objective. And so on. In my diaries I named this organization The Powers.

All of this, because I was 100% sure I would succeed at everything. Whenever I woke up in a psych ward, I just chalked it up to the will of The Powers. Everything I experienced was filtered through my lens of success—which nothing could contradict. That is, until one day, a few years ago, I woke up, acquiring the highest level of relative sanity in years.

To be continued, later in this chapter, in Section 14.7.

14.3 Are you 100% sure you will succeed at everything?

If you are 100% sure you will succeed at everything, I suggest playing hard to get. You don't really need to look hard for secret messages. If you really are that important, and if The Powers really want to transmit a message to you, make them work for it. If they really are that powerful, and if they really want to give you a message, it is definitely within their power and interest to deliver an unambiguous message to you.

For example, suppose you are receiving messages through ESP: don't settle for a subtle hand gesture to indicate you should go down a rabbit hole. Post something on the Web, or wherever else The Powers can read your writing (and other people won't). Say to The Powers: "If you want me to go down a rabbit hole, use ███████████ as a code word to send me down a rabbit hole."

The point here is that if you really are that important, and if The Powers really are that powerful, and if they really want you to go down a rabbit hole—they will use your code word. In general, if you play hard to get you will receive unambiguous messages.

In the film *The Matrix 4 (Resurrections)*, The Powers are the humans outside the Matrix trying to help Neo escape the Matrix. They go to great lengths to persuade Neo to go down their rabbit hole, even laboring to hack a mirror so Neo will know they are for real, and Neo should join them by taking the red pill.

14.4 Or, are you only X% sure?

OK. Suppose you estimate there's an X% chance your life is extraordinary, and a Y% chance your life is ordinary. What now? I claim you don't need to go all-in on your extraordinary narrative. Rather, I suggest guiding the evolution of your belief system towards *mutual success*—i.e., succeeding regardless of which narrative turns out to be more true.

It works like magician's rationality, from Chapter 9. To review, if Alice and Bob are both magicians, then they both believe they will succeed, and each other will succeed. Therefore, it would be illogical to behave in a way that prevents either of their successes. Applying this concept to (a) your extraordinary narrative and (b) your ordinary narrative, then simply refrain from behaving in a way that would thwart either of their successes.

For example, in the process of writing this book, I sporadically experienced visions of grandeur. Specifically, at certain times, I felt almost certain that my book would be a smashing success, and would help rapidly usher in world peace, peacefully. I believed penning this book was a personal rabbit hole that The Powers really do appreciate. I have intentionally guided the evolution of my belief system towards taking risk-averse cues from The Powers, and have learned from my mistakes. Risky rabbit holes have stung myself and others in the past, so I avoid them.

So how does writing this book achieve mutual success? Well, if the book turns out to be extraordinary, I have definitely succeeded. But if the book is ordinary, I have succeeded too. Writing it has been fun, I've learned at least a few lessons, and I might even entertain a few friends and family members, too. I win either way. This book is neither a risky investment for my extraordinary narrative nor my ordinary narrative.

14.5 Reality distortion fields

Let's now suppose you're a magician, but you don't believe in yourself a tremendous amount. Thus you would like to boost your faith in yourself, to boost your actual success. You might be tempted to develop a "reality distortion field."

According to legend, the members of Steve Jobs' inner circle believed he possessed a *reality distortion field*. It was their pejorative for Steve Jobs' ability to live in an imaginary future, while enticing people to join into his imaginary reality, which created the conditions for Steve Jobs' imaginary vision to become objective reality. Cool, right?

According to the legend, Steve Jobs fired employees who wouldn't indulge his reality distortion field. Maybe not cool.

So, how can you create your own reality distortion field? Well, if your confidence in your own success is shaky, you can simply refrain from associating with skeptics. This way, you're only surrounded by cheerleaders, which might boost your belief in your own success, boosting the actual likelihood of success. It might also make you lonely, and it might also lower your chance of success.

How? Well, what if you enthusiastically climb up the wrong ladder? You're only surrounded by cheerleaders, and you believe in yourself, so how would you know you're up the wrong ladder? According to magician, and supposed expert on life, Joseph Campbell:

> There is perhaps nothing worse than reaching the top of the ladder and discovering that you're on the wrong wall.

In scientific terms, over relying on cheerleading and the secret to success can be a problem because of the *confirmation bias*.

14.6 Confirmation bias

Humans, generally speaking, prefer the beliefs they currently hold, regardless if those beliefs are accurate or mistaken. The preference for your current beliefs results in the confirmation bias. The confirmation bias is the bad habit of only seeking and considering evidence that conforms to your beliefs. The bad news is the confirmation bias is pervasive amongst humans. The good news is that you shouldn't be embarrassed; it's pervasive.

If you want to suss out mistaken beliefs, you should subject your beliefs to skepticism. Balancing belief and skepticism is an art that I believe is essential to success. And, if you're unwilling to subject your beliefs to skepticism, you might just march up the wrong ladder.

14.7 The origins of the Zen lemonade garden, Part 2

This is a good spot to resume my story, the story of the genesis of the Zen lemonade garden. When my mind popped, all those years ago, I had just left professional skepticism six months prior, and had dove into magic and Buddhism. With a popped mind, I suddenly made a connection between Zen, koans, doormen, and bouncers. Specifically, I realized Zen masters could use doormen, bouncers, and koans to ward off antagonists!

A simple, beautiful, elegant proof revealed itself to me, a proof for the Zen lemonade garden. I rushed to the computer, and typed it off and sent it to a handful of my former collaborators. They told me it was gibberish, and they were worried about me. When I protested, they said my proof was woefully tautological—an obvious blunder no professional mathematician would ever make. I scoffed them off. I was certain of my success.

Off and on for years, I would sporadically try to re-prove the Zen lemonade garden during bouts of delusions of grandeur. But the confirmation bias haunted me; my resolute belief in my own success thwarted my previous ability to evaluate my own work. It's stunning how powerful the confirmation bias is when combined with certainty of success.

During that time, during my bouts of delusions, I witnessed real magic. And now, having gained stronger footing into reality, I am still convinced much of the magic I have witnessed is real.

In developing this book, I once again set out to prove the Zen lemonade garden. This time, though, I held the confirmation bias acutely in mind. I failed. But, since I'm a magician, I tried again, and failed, and then saw yet another glimmer of hope—pursued it, discovered $w = 1 + \log_2(1/t)$ and suspected I was about to succeed. Then, I actually succeeded.

I succeeded! I finally did it; after near a decade of trying, I have proved the Zen lemonade garden via simulation. And may it contribute to world peace! Thank you skepticism! Thank you magic! Thank you Buddhism! Thank you to all the magicians and skeptics who have helped me! Thank you, thank you, thank you!

14.8 Conclusion

A word of inspiration from Joseph Campbell:

> If you follow your bliss, doors will open for you that wouldn't have opened for anyone else.

Lastly, in the words of artist Barbara Kruger:

> Belief + Doubt = Sanity

Chapter 15

The wish giving gem

According to legend, if you dig deep enough in your backyard, you will find a wish-giving gem. You will receive one million wishes! Sadly though, most of your wishes won't come true. This gem is not a wish-*fulfilling* gem; this gem is a wish-*giving* gem. It gives you wishes, but most of them do not come true.

Many people consider the wish giving gem to be a curse. For many fall into sadness when their wishes don't come true. When little Kata dug up the wish giving gem in her backyard, her parents scolded her. Her parent's said, "Well, now that you're filled with one million wishes, you're cursed!"

Kata wished for the wishes to go away. But, the wishes remained, which made her sad. She asked her parents what to do. They said, "When we became cursed, when we were children, we asked all the other cursed people what to do. You should do the same."

She listened to her parents. Whenever Kata asked someone about the curse, they would frown, then offer advice. Kata heard:

- "Wish for other people to have wishes," and

- "Bury the gem back into the Earth," and

- "Don't make selfish wishes," and

- "Concentrate all your wishes into one big wish," and
- "Wish for money," and
- "Wish for something different every time," and
- "Get used to disappointment."

None of the advice seemed to help. Sure, some of her wishes came true. But, she was still sad for all her wishes that didn't come true.

Kata's life became miserable. One day, she began crying in class. Her teacher frowned, then smiled, and said, "Kata, would you like to go to the library?"

So, she went to the library. She sat down, and cried. The librarian walked to Kata, frowned, then smiled, and said, "You're crying because of the wish giving gem, aren't you?"

Kata's eyes grew wide. She asked, "How did you know?" The librarian smiled, and said, *"I know the secret of the wish-giving gem."*

Kata jumped up! She said, "Oh please tell me!" The librarian smiled and said, "I know the secret, but I can't explain it. I do know though where the books on wishes are."

The librarian pointed, and Kata ran off. There were books and books and books on wishing. One by one, Kata opened each book.

Most of the books had strange words everywhere, words like "obfuscate," and "unintelligible." Kata kept searching.

Eventually, she found a book that made sense, *The Art of Wishing*. The book contained 21 words total, 7 pictures total, and 3 punctuation marks total. One word was underlined. The words read:

> *Wishes don't make people sad; <u>clinging</u> to wishes makes people sad. Hold wishes lightly in your heart as you dream big.*

I would be delighted to tell you that Kata lived perfectly happily ever after. However, it's not that simple.

Yes, Kata learned the secret of the art of wish making, but skillful wish making takes lots of practice, lots of effort, and lots of wishes.

But even though Kata now knew the secret, no one else believed her. Whenever she told people about her book, they said she was being "unintelligible." Whenever she showed people the book, they said the book was "obfuscated."

The years rolled by, and Kata kept practicing. And, Kata did eventually become more skillful at wish making. ☺

Years later, one day, someone asked her for advice about a gem they had dug up in their backyard. She frowned, then smiled, and said, "There's a wonderful section on gems at the library." She pointed her finger, and the child ran off.

So, as I said, Kata didn't exactly live perfectly happily ever after. But, in the end, she was happy to have discovered the wish giving gem. The end.

Part II

Appendix: Magician's rationality

Chapter 16

Rationality

To give a sneak preview, a few chapters ahead, in Chapter 9, I present *magician's rationality*—a magical strategy that guarantees success in certain circumstances, under certain mathematical assumptions. Thus, I must cover preliminary material. Fortunately, this material offers a frightening story from a dystopic future, so enjoy!

We begin with the classic concept of *rationality*, from the mathematical field of game theory.

16.1 Rationality

In game theory, a player is *rational* if they always make the best decisions towards achieving their objective. Here, rationality is a mathematical concept, that only applies to mathematical models of games. From my anecdotal survey, it seems that game theory is focused on achieving selfish objectives, and thus rationality is usually synonymous with selfishness. Thus, in keeping with tradition, I simply use the term rationality to refer to selfish rationality.

Rationality has its problems. Among them: sometimes rationality prescribes noncooperation, even when mutual cooperation would have been better. The classic way to introduce this problem is to explain the *prisoner's dilemma*.

16.2 The prisoner's dilemma

Suppose you live in a dystopic future, and you are walking to work one day. Suddenly you are snatched up by the police, for no good reason. They take you to an interrogation room, and one of the interrogators says: "We want you to snitch on your accomplice."

You say, "I don't have an accomplice. I'm innocent."

They say, "We don't care. We want you to snitch anyway."

"Why should I snitch?"

"We'll offer you a deal."

"What's the deal?"

They say, "If you snitch, and your accomplice doesn't, you go free."

"And what if I don't?"

"If you *don't* snitch, but they do, then you go to prison for three years."

Yikes. So far snitching sounds pretty good. But questions remain in your mind.

You ask, "What if we both snitch?"

"Each of you serves two years in prison."

Hmmm.... that doesn't sound too good.

You ask your last question, "What if neither of us snitch?"

"Then you each go to prison for one year."

That's the prisoner's dilemma. So, what should you do?

16.3 Game theory

According to the field of *game theory*, there are various solutions. The dilemma was first conceptualized in 1950 by a pair of mathematicians, to help figure out nuclear-warfare strategy for the United States. According to them, the "rational" solution is to snitch. Thus when two "rational" players play against each other, they both snitch, and they both end up in prison for two-years, each.

Wouldn't it be nice if it were "rational" to not snitch? Then, if two "rational" players played against each other,

16.4. APPLICABILITY TO THE REAL WORLD 115

neither would snitch, and they would each only serve one year in prison.

What is the game theoretically rational choice to achieve the best outcome for yourself? The answer is snitching. And if your "accomplice" is rational too, then they snitch and you each go away for *two years*. This result is unfortunate, because if you had both refrained from snitching, you both would have only served *one year*, each.

So, why is snitching rational? The rational thought process goes like this.

1. What are my "accomplice's" options?

    a) They snitch

    b) They don't snitch

2. What should I do in either scenario?

    a) If they snitch:

    　　i. If I snitch, I serve two years

    　　ii. If I don't snitch, I serve three years

    　　iii. Therefore: *if they snitch, I should snitch*

    b) They don't snitch

    　　i. If I snitch, I serve zero years

    　　ii. If I don't snitch, I serve one year

    　　iii. Therefore: *if they don't snitch, I should snitch*

3. Therefore, in either scenario, my best option is to snitch

Therefore, you snitch. And if your "accomplice" is rational, too, then they snitch, and you each go to prison for two years. Wouldn't it be nice if it were rational to cooperate by not snitching?

16.4 Applicability to the real world

We have established that mathematical rationality results in noncooperation, for at least one mathematical game, even when mutual cooperation would have offered a better result for everyone. Perhaps game theory models reality well. Or, perhaps this result illustrates that game theory fails to model reality. As in, perhaps, in the real world, people are not mathematically rational—rather, perhaps people succeed at cooperation, even when math says its "irrational."

Thankfully, we don't need to speculate. Scientists have conducted experiments with actual humans, to see how they behave in dilemmas similar to the prisoner's dilemma. For example, in one experiment conducted inside a prison, with actual prisoners for players, with coffee and cigarettes at stake, prisoners cooperated 56% of the time. In another experiment, conducted as part of the same study, 37% of college students cooperated, with money at stake.

Thus, game theory's model of rationality doesn't always match reality so well.

16.5 Bibliograhpy

[1] Menusch Khadjavi &Andreas Lange, "Prisoners and their dilemma." Journal of Economic Behavior & Organization, Volume 92, August 2013, Pages 163-175.
https://www.sciencedirect.com/science/article/abs/pii/S0167268113001522

[2] Rose Eveleth, "What Happens When You Test the Prisoner's Dilemma on Prisoners." Smithsonian Magazine, July 25, 2013.
https://www.smithsonianmag.com/smart-news/what-happens-when-you-test-the-prisoners-dilemma-on-prisoners-18221040/

Chapter 17

Superrationality

Game theorists have set out to discover new notions of rationality. Perhaps they're perturbed by game theory's classical notion of rationality. Perhaps there exist models of rationality that match reality better; perhaps such models could explain cooperative behavior that defies classical rationality. And, perhaps if we had a better model of rationality, it could inspire more cooperative behavior.

One of the most celebrated non-classical models of rationality is *superrationality*—which succeeds in overcoming the prisoner's dilemma, in certain situations.

In Douglas Hofstadter's book *Metamagical Themas*, Hofstadter conceptualizes superrationality. It is a notion of rationality that seems better than the classical notion of rationality in many circumstances, if only its assumptions were realistic. According to Hofstadter's model: (1) superrational players assume their partners are also superrational players, and (2) superrational players are classically rational, except *all superrational players are guaranteed to use the same strategy.*

Let's see how superrationality fares in the prisoner's dilemma.

17.1 Superrational prisoner's dilemma

Once again, you're in dystopia, you are abducted and offered the following deal. But this time, you happen to be superrational. To review, here is the deal the interrogators offer you:

1. If both you and your "accomplice" snitches, then you each serve two years

2. If neither you, nor your accomplice snitches, then you each serve one year

3. If you snitch, and your accomplice doesn't, then you go free, and your accomplice serves three years

4. If you don't snitch, but you accomplice does, then you serve three years, and your accomplice goes free

So what do you, a superrational player, do? First, you assume your accomplice is superrational. Since you and your accomplice are guaranteed to use the same strategy (as you believe), you can eliminate two possibilities: the ones where you and your accomplice choose different strategies.

1. If both you and your "accomplice" snitches, then you each serve two years

2. If neither you, nor your accomplice snitches, then you each serve one year

3. ~~If you snitch, and your accomplice doesn't, then you go free, and your accomplice serves three years~~

4. ~~If you don't snitch, but you accomplice does, then you serve three years, and your accomplice goes free~~

Then, you choose the best of the remaining two options, which is to not snitch. If the other player turns out to be superrational, then job well done: now you're only serving one year in prison.

17.2 Critique of superrationality

I think superrationality is super cool, because it empowers cooperation. However, for superrationality to work in the real world, you must pair up with someone who happens to be superrational too—or at least someone who decides to cooperate. If you pair up with a classically rational player (or anyone else who snitches), then you go to prison for three years. Thus, for real-world applicability, we need to ask: how prevalent is superrational thinking in practice? I do not know, and am unaware of any scientific studies.

Further, there are other issues with superrationality. Consider the *gambler's dilemma*.

Definition 6. *The gambler's dilemma*

The gambler's dilemma illustrates how superrationality might not lead to "success." The game goes like this.:

- If you both cooperate, you each win $2.
- Or, you both defect, you each win $1.
- But if one of you cooperates, and the other defects, the defector wins $100 and the cooperator wins $0.

According to superrationality, you should narrow your choices down to either both cooperating, or both defecting. In this case, you would choose a small victory of $2 each. Perhaps that is not as cool as a chance at a hundo though. If both players make their choice via a coin toss, each player's expected outcome goes up.

17.3 Bibliograhpy

[1] Douglas Hofstadter, *Metamagical Themas*, 1985.

Chapter 18

Hyperrationality

Chapter 19

Misc

In Chapter 9, I presented a formal proof of magician's rationality, but skipped out on the finer details, and also skipped out on the philosophical underpinnings of my model. In this appendix, I re-present the proof, but this time with full, formal details—for those interested in such a thing.

But before we dive into the proof, I would like to discuss the philosophy (or rather, philosophies) of belief, which this chapter is about.

19.1 Degrees of belief

19.2 Beliefs and behavior

19.3 Knowledge vs. beliefs

19.4 Strong vs. weak sss

19.5 IFS

Bountiful research suggests individuals aren't really individuals. Or rather, it can be helpful to think of an individual (to model an individual) as a collection of egos, who take turns assuming, controlling the body—each with their own beliefs, opinions, desires, inclinations, prejudices, and so on.

19.6 Sss is magic, magician's rationality is reduced to logic

Chapter 20

Introduction

In Chapter 9 we informally explored magician's rationality. To recap, I claimed if you *really* belief something, then you behave according to that belief. But, you also need to be logical enough to connect your beliefs to behaviors that are compatible with your beliefs.

Then, I framed the supposed secret to success in terms of logic, behavior, and deep belief. Essentially, even if your belief in success is *small*, so long as you *deeply* believe in the supposed secret, then your logic will conclude a new belief: a *deep* belief in success. Intuitively, the supposed secret to success is like a machine, located in your brain. The machine takes a belief of success (even a little belief), cranks some gears, then spits out a deep belief in success. While this jump to a big belief might help you succeed via magic, this jump is not sufficient to *logically guarantee* success. However, once you and your fellow magicians have made the jump altogether, then magician's rationality comes into play—logically guaranteeing success under certain conditions.

Here's the gist of how magician's rationality works. Every magician (i.e. every sufficiently logical person who deeply believes in the supposed secret to success) ought to believe every magician will succeed. This insight is the foundation of magician's rationality. I think the logic here is pretty straightforward. Essentially, it's just the recognition that the supposed secret to success works for other magicians, the same way it works for you. If you are a magician, and your friend Bob is a magician, then it is logical to conclude Bob will succeed. Similarly, Bob will conclude you will succeed.

Going further, if a group of people are all magicians, and each believes the others are magicians, then everyone believes in everyone's success, then they will behave accordingly—they will behave in a manner that is consistent with everyone's success, and they will not behave in a manner that would prevent anyone's success. Thus, success is logically guaranteed—under certain conditions

Of course, this is just a logical model of reality (See Section 2.2). However, I believe it's a good model; it reflects my anecdotal research—my personal experiences as a magician. This chapter dives into this model, formally. Thus, we can rigorously inspect my logic, as well as understand all the caveats that might thwart the guarantee of success.

20.1 Outline

... Chapter 21 begins with a formal model of what it means to be a "sufficiently logical human."...

Chapter 21

Sufficiently logical humans

21.1 Introduction

This chapter presents a formal model of what it means to be a "sufficiently logical human." Essentially, we model humans as computers that attempt to make logical decisions. However, these computers are imperfect. Sometimes they fail to realize interesting logical truths. Thus a sufficiently logical human (i.e. a sufficiently powerful computer) is "sufficient" when it is logical enough realize interesting truths.

21.2 First-order logic

In this chapter, we model sufficiently logical humans using first order logic.

21.3 Shallow beliefs

A *shallow belief* is a formula in first-order logic—a formula that is regarded as true, without needing proof. For example, if Alice assumes $x > 3$, then the formula $x > 3$ is a part of Alice's "belief system."

21.4 Belief systems

In our model each human believes zero or more shallow beliefs, which are wrapped in a personal belief system. Formally, a belief system Φ is an ordered list of beliefs, arranged from most-believed to least-believed belief. Continuing our example from the previous section, $x > 3 \in \Phi_{\text{Alice}}$. These symbols simply say, "$x > 3$ is in Alice's belief system."

21.5 Operations on belief systems

For a given belief system Φ, the notation $\Phi.\text{top}$ refers to the most-believed belief in the belief system Φ. Relatedly, the notation $\Phi.\text{lower}$ refers to the belief system Φ, but with the most-believed-belief removed.

Also, the notation $\Phi = [a, b, c]$ means that a, b, and c are beliefs in the belief system. And furthermore, putting it all together:

$$\Phi = [a, b, c] \tag{21.1}$$
$$\Phi.\text{top} = a \tag{21.2}$$
$$\Phi.\text{lower} = [b, c] \tag{21.3}$$

21.6 Provable from

In first-order logic, the symbol $\vdash$ means "provable from." For example, suppose Bob's belief system contains two formulas $\Phi_{\text{Bob}} = [a, b]$. Becoming more concrete, suppose:

$$a = (x \geq 7) \tag{21.4}$$
$$b = (x \leq 9) \tag{21.5}$$

Clearly, Bob can conclude:

$$\varphi = 7 \leq x \leq 9 \tag{21.6}$$

Essentially, what this means is c is provable from Φ_{Bob}. To use our fancy symbol, we could just write:

$$\Phi_{\text{Bob}} \vdash \varphi \tag{21.7}$$

21.7 Introduction to deep beliefs

In our model, a "deep belief" is a formula that is *relatively* true according to a belief system, yet might not be a shallow belief in the belief system. For example, in the preceding section φ *might* be a deep belief according to Bob's belief system, since $\Phi_{Bob} \vdash \varphi$.

To explain the notion of *relatively true*, and why φ *might* be a deep belief, and might not, we must discuss *automated theorem provers*.

21.8 Automated theorem provers

In our model, an automated theorem prover is an algorithm that takes a belief system Φ, and and tries to discover deep beliefs that are logically implied by the belief system.

21.9 Personal theorem provers

In our model, we assume every human possesses an automated theorem prover—their *personal theorem prover*. However, these theorem provers are imperfect, which represents the reality that human reasoning is limited.

So, even though $\Phi_{Bob} \vdash \varphi$, perhaps Bob is not logical enough to realize φ is a logical conclusion from a and b.

Therefore, to represent imperfect theorem provers, we use the symbol $\models_{Bob}$ to denote "provable according to Bob's theorem prover." Thus, $\Phi_{Bob} \models_{Bob} \varphi$, means: "Bob can prove formula φ is true, according to his belief system." To shorten the notation even further, we can omit Bob's name from the "provable from" operator. Thus, we can simply write: $\Phi_{Bob} \models \varphi$, and it is implied that φ is provable from Bob's theorem prover and belief system.

21.10 Definition of deep beliefs

A *deep belief* is a formula in first-order logic that is provable by a particular theorem prover, given a particular belief system. I.e. if $\Phi_{\text{Bob}} \models \varphi$, then φ is a deep belief for Bob.

Some shallow beliefs become deep beliefs, but not always, because shallow beliefs might contradict each other, which brings us to the next section.

21.11 Algebraic manipulation of deep beliefs

Personal theorem provers need to work hard to arrive at deep beliefs. Thus you cannot simply manipulate deep beliefs with algebra. For example, suppose $\Phi \models x = 1$ and $\Phi \models y = 2$. The theorem prove might not necessarily prove $\Phi \models x = 1 \land y = 2$. The only way one can arrive at a conclusion regarding a deep belief is either by direct assumption (i.e. "we assume $\Phi \models x = 1 \land y = 2$") or an assumed implication the person is capable of making (i.e. "we assume Alice is sufficiently logical to realize: $[\Phi \models \varphi_1] \land [\Phi \models \varphi_2] \implies \Phi \models \varphi_1 \land \varphi_2$").

21.12 Contradictory beliefs

Belief systems may contain contradictory beliefs. For example, Φ_{Alice} might equal $[a, b]$, where:

$$a > 0 \tag{21.8}$$
$$a < 0 \tag{21.9}$$

This is not a problem for our personal theorem provers, because they ignore contradictory beliefs—if and when contradictions are detected. To be more specific, if a prover detects a contradiction, the prover ignores the less believed beliefs which led to the contradiction.

Let's consider a concrete example. Suppose Alice's belief system contains three beliefs $\Phi_{\text{Alice}} = [a, b, c]$. Becoming more concrete, suppose:

$$a = (x \geq 7) \qquad (21.10)$$
$$b = (x \leq 9) \qquad (21.11)$$
$$c = (x < 0) \qquad (21.12)$$
$$\qquad (21.13)$$

If Alice's prover is powerful enough, it will notice a and c contradict each other. Thus, it will ignore c and could prove $\varphi = 7 \leq x \leq 9$ (if it's powerful enough). Thus, $\Phi_{\text{Alice}} \models \varphi$ is possible, depending on the power of Alice's prover.

21.13 From personally provable to provable

Every personally provable formula is also a provable formula, but not the other way around. Formally:

$$\Phi \models \varphi \implies \Phi \vdash \varphi \qquad (21.14)$$
$$\neg(\Phi \vdash \varphi \implies \Phi \models \varphi) \qquad (21.15)$$

21.14 Sufficiently logical humans

Now for the definition of a "sufficiently logical human." First, it important to point out that, in our model, the notion of sufficiently logical is always relative to a theorem. As in, "Alice is sufficiently logical to prove such and such a theorem." What is a "theorem" in our model? A *theorem* here is a deep belief that is a logical conclusion from one or more beliefs (either shallow or deep), via a perfect theorem prover. Let's just

CHAPTER 21. SUFFICIENTLY LOGICAL HUMANS

dive into the formal definition, before looking at it informally.

$$\text{sufficiently-logical(} \tag{21.16}$$
$$s_1, \tag{21.17}$$
$$s_2, \tag{21.18}$$
$$s_3,, ... \tag{21.19}$$
$$\varphi_1, \tag{21.20}$$
$$\varphi_2, \tag{21.21}$$
$$\varphi_3, ... \tag{21.22}$$
$$\phi, \tag{21.23}$$
$$\Phi, \tag{21.24}$$
$$\models) \tag{21.25}$$
$$= [[\Phi \supseteq \{s1, ...\} \land ... \land (\Phi \models \varphi_1) \land ...] \implies (\Phi \vdash \phi)] \implies \Phi \models \phi \tag{21.26}$$

To reveal how this is a reasonable definition of "sufficiently logical," let's tentatively assume that everyone is sufficiently logical. That means if a theorem is provable by a perfect theorem prover, then it is provable by a personal theorem prover. That's exactly when this boolean function returns true. OK, now stop assuming everyone is sufficiently logical. 😊

21.15 Conclusion

TODO

Chapter 22

Belief-wise games

22.1 Introduction

Informally, a *belief-wise game* is any game similar to the prisoner's dilemma, but with room for more than two players, plus beliefs, plus conversations, plus varying degrees of logical-thinking prowess. With regards to this book, belief-wise games are important, because our model of magician's rationality only applies to belief-wise games (see the following chapter).

In the game, there are N players. Each player has one choice to make: to either "cooperate" or "defect," all players cast their ballots (their choice) in secret, simultaneously, with each player's identity (e.g. unique name) attached to their ballot. Once all ballots have been cast, the game awards victory points to each player (possibly negative), as a function of the ballots.

22.2 Ballots

Each player i casts a ballot $b_i \in \{C, D\}$.

22.3 Victory points

Let v_i represent the victory points awarded to Player i, at the conclusion of the game.

22.4 The **game** function

The game logic is represented as a function named **game** that maps final ballots to final scores. Formally:

$$\mathsf{game} : \{C, D\}^N \to \mathbb{R}^N \qquad (22.1)$$

For example, consider one outcome from the prisoner's dilemma:

$$b_1 = C \qquad\qquad b_2 = D \qquad (22.2)$$
$$\mathsf{game}(b_1, b_2) = (v_1, v_2) \qquad (22.3)$$
$$v_1 = -3 \qquad\qquad v_2 = 0 \qquad (22.4)$$

But, because we use first-order logic, we must encode our **game** function in a format compatible with first-order logic. To do so, we permute over the ballot possibilities, to define 2^N implications of the form:

$$(b_1 = C \wedge b_2 = C \wedge ...) \implies \qquad (22.5)$$
$$(v_1 = \mathsf{game}(C, C, ...)[0] \wedge v_1 = \mathsf{game}(C, C, ...)[1]...) \qquad (22.6)$$
$$... \qquad (22.7)$$

22.5 Success

Each Player i defines their own personal success criteria, as a minimal number for v_i. And, players also define success criteria for other players. That's right. Alice can determine whether or not Bob "succeeded," depending solely upon Alice's criteria for the success of Bob.

Formally, success criteria is expressed as follows. Let $s_{i,j}$ be the minimum number of victory points that Player i must win to be successful, according to Player j.

All players are successful, according to their personal criteria, if: $v_1 \geq s_{1,1} \wedge v_2 \geq s_{2,2} \wedge ... \wedge v_N \geq S_{N,N}$

22.6 Conversation

Belief-wise games allow players to meet in person, and have conversations. Conversations allow players to acquire beliefs about the other players' beliefs, logical capabilities, and success criteria—before casting ballots. In this model, there is no time limit to cast ballots, though we assume all players eventually cast their ballots. Players may of course lie and be deceptive during conversations.

During conversation, players update their beliefs and success criteria. At the conclusion of the conversation, but before ballots are cast, we assume every player has finalized their belief system, Φ_i. We previously discussed belief systems in Section 21.4.

Once everyone's deep belief system is finalized, and success criteria are finalized, then each player casts their ballot, victory points are computed, and the game ends.

22.7 Belief notation

Using first-order logic, the expression $\mathcal{B}_{\text{Alice}}(x = \mathit{formula})$ indicates Alice believes x to be true, to some degree. Thus, $\mathcal{B}_{\text{Alice}}(x)$ is merely shorthand for $x \in \Phi_{\text{Alice}}$. Remember, beliefs can be contradictory (see Section 21.12).

22.8 Example belief: $\mathcal{B}_i(v_j = 5)$

Humans might hold beliefs about who will earn which victory points. For example, $\mathcal{B}_i(v_j = 5)$, means Player i believes Player j will have 5 victory points at the conclusion of the game.

22.9 Players accurately understand the game

At the root of every Player's belief system Φ_i, players hold accurate beliefs about the game. I.e. the **top**-most beliefs include the **game** function and $b_i \in \{C, D\}$.

22.10 Relatively logical behavior

Recall our philosophical discussion of what it means to *really* believe something (Section **??**). Basically, behavior is necessarily consistent with one's deeply held beliefs. Thus, in a belief-wise game, a player may only cast a ballot that is consistent with their beliefs.[1]

Formally, Player i may only cast the ballot b_i if it is consistent with their beliefs. A ballot choice $b_i = x$ is consistent with Player i's beliefs if it does not contradict Φ_i according to Player i's personal theorem prover, which has limited logical prowess. Formally:

$$\textsf{consistent-ballot}_i(x) \equiv \neg(\Phi_i \models b_i \neq x) \qquad (22.8)$$

If only one ballot choice is consistent with a player's belief system, the player makes that choice. If two choices are available to a player, the player arbitrarily chooses one choice.

What about if the player cannot make a logical choice? That's actually not possible. Because player's deeply believe $b_i \in \{C, D\}$ (see the previous section), this means players are always able to vote. In the worst case scenario, where a player adopts beliefs that are contradictory to voting relatively logically, those beliefs are discarded, and the player makes an arbitrary choice for their ballot.

Here is the decision making procedure:

---

[1] And remember, when shallower beliefs contradict deeper beliefs, the shallower beliefs are discarded, thus achieving a notion of consistency and a notion of what it means for a belief to be "deeply" held. See Section **??**.

$$\neg\text{consistent-ballot}_i(D) \implies b_i = C \quad (22.9)$$
$$\neg\text{consistent-ballot}_i(C) \implies b_i = D \quad (22.10)$$
$$\text{consistent-ballot}_i(C) \wedge \text{consistent-ballot}_i(D) \implies b_i \in \{C, D\} \quad (22.11)$$

22.11 Beliefs about beliefs

Players might believe other players hold certain beliefs. For example: $\mathcal{B}_i(\mathcal{B}_j(s_{i,i} = 7))$ means Player i believes Player j believes Player i wishes to earn at least 7 victory points.

Recalling from Section ??, the definition of $\mathcal{B}$ notation, the following expressions are equivalent:

$$= \mathcal{B}_i(\mathcal{B}_j(s_{i,i} = 7)) \quad (22.12)$$
$$= ((s_{i,i} = 7) \in \Phi_j) \in \Phi_i \quad (22.13)$$

22.12 Hoisting

We can *hoist* variables into nested scopes, using the $\leftarrow$ operator. Consider the following belief:

$$\mathcal{B}_j(v_j \geq x)_{x \leftarrow 5} \quad (22.14)$$
$$= \mathcal{B}_j(v_j \geq 5) \quad (22.15)$$

Consider another example:

$$\mathcal{B}_i(\mathcal{B}_j(v_j \geq x)_{x \leftarrow s_{j,i}}) \quad (22.16)$$

Here, Player i believes something about Player j's beliefs. In particular Player i believes Player j believes they will succeed, according to *Player i's actual success criteria for j*.

22.13 Never confused about own success criteria

A player's belief is always accurate about their personal success criteria. E.g.:

$$\mathcal{B}_i(s_{i,i} > 5) \equiv \mathcal{B}_i(x > 5)_{x \leftarrow s_{i,i}} \qquad (22.17)$$

Similarly, player's are never confused about their criteria for other players. E.g.:

$$\mathcal{B}_i(s_{k,i} > 5) \equiv \mathcal{B}_i(x > 5)_{x \leftarrow s_{k,i}} \qquad (22.18)$$

22.14 Never confused about own beliefs

Players are never confused about their own beliefs. When a player believes believes something, then they believe they believe that thing. Conversely, when a player believes they believe something, then they believe that thing. Formally,

$$\mathcal{B}_i(x) \equiv \mathcal{B}_i(\mathcal{B}_i(x)) \qquad \mathcal{B}_i(x) \equiv \mathcal{B}_i(\mathcal{B}_i(x)) \qquad (22.19)$$

22.15 Conclusion

Chapter 23

Theory of magician's rationality

23.1 Introduction

23.2 The supposed secret to success

Let's build the formula for the supposed secret to success, as applicable to belief-wise games. First, we express that Player j shallowly believes they will succeed according to their personal definition of success.

$$\mathcal{B}_j(v_j \geq s_{j,j}) \tag{23.1}$$

Next, we express that Player i believes Player j will succeed (according to Player j's personal definition of success, whatever Player j happens to believe success is made of).

$$\mathcal{B}_i(\mathcal{B}_j(v_j \geq s_{j,j})) \tag{23.2}$$

Now we twist it a little. We express that Player i believes Player j will succeed, but this time according to Player i's criteria for j. This expression actually expresses *Player i believes Player j will succeed*.

$$\mathcal{B}_i(\mathcal{B}_j(v_j \geq x)_{x \leftarrow s_{j,i}}) \tag{23.3}$$

Let's dive into this a little more concretely. Suppose $s_{j,i} = 5$. I.e, Player i considers Player j successful, if Player j scores at least 5 points.

Since Player i is not confused about their criteria for j (see Section 22.13), then, in this case, 23.3 is equivalent to:

$$\mathcal{B}_i(\mathcal{B}_j(v \geq 5)) \qquad (23.4)$$

Moving along towards a definition of the supposed secret to success. Now we express that Player i believes Player j will succeed (according to Player i's actual criteria), so long as Player j believes they will succeed (according to Player i's actual criteria).

$$\mathcal{B}_i(\mathcal{B}_j(v_j \geq x)_{x \leftarrow s_{j,i}} \implies v_j \geq s_{j,i}) \qquad (23.5)$$

Since this belief about Player j applies to all possible j's, we simply apply the belief to all j. This expression is the supposed secret to success:

$$\mathcal{B}_i(\forall j.[\mathcal{B}_j(v_j \geq x)_{x \leftarrow s_{j,i}} \implies v_j \geq S_{j,i}]) \qquad (23.6)$$

As a shorthand, the supposed secret to success can be referred to as sss. For example, to express that Player i believes in the supposed secret to success, we could just write:

$$\mathcal{B}_i(sss_i) \qquad (23.7)$$

Observe, every belief in $\mathcal{B}_i(sss_i)$ is a shallow belief.

23.3 Personal implication of the supposed secret to success

Now that we have a definition of the supposed secret to success, let's prove some thing.

23.3. PERSONAL IMPLICATION OF THE SUPPOSED SECRET TO SUCCESS

Lemma 2. *If (Condition 1) Player i at least shallowly believes they will succeed, (Condition 2) if they deeply believe in the supposed secret to success, and (Condition 3) if they are sufficiently logical, then (Conclusion) the player will deeply believe they will succeed.*

Before presenting the proof, let's begin by formalizing the lemma.

$$\text{(Condition 1)} = \mathcal{B}_i(v_i \geq s_{i,i}) \tag{23.8}$$

$$\text{(Condition 2)} = \Phi_i \models_i sss_i \tag{23.9}$$

$$\text{(Condition 3)} = \textsf{sufficiently-logical(} \tag{23.10}$$

$$s_1 = v_i \geq s_{i,i}, \tag{23.11}$$

$$\varphi_1 = sss_i, \tag{23.12}$$

$$\phi = v_i \geq s_{i,i} \tag{23.13}$$

$$\Phi = \Phi_i \tag{23.14}$$

$$\models \ = \ \models_i) \text{ and} \tag{23.15}$$

$$\text{(Conclusion)} = \Phi_i \models_i v_i \geq s_{i,i} \tag{23.16}$$

Proof. We begin by assuming all three conditions are true. Next we review the definition of **sufficiently-logical**(...):

sufficiently-logical((23.17)
$$s_1,$$ (23.18)
$$s_2,$$ (23.19)
$$s_3,,\ldots$$ (23.20)
$$\varphi_1,$$ (23.21)
$$\varphi_2,$$ (23.22)
$$\varphi_3,\ldots$$ (23.23)
$$\phi,$$ (23.24)
$$\Phi,$$ (23.25)
$$\models)$$ (23.26)

$$= [[\Phi \supseteq \{s1,\ldots\} \wedge \ldots \wedge (\Phi \models \varphi_1) \wedge \ldots] \implies (\Phi \vdash \phi)] \implies \Phi \models \phi$$
(23.27)

Applying (Condition 3) to the definition of sufficiently-logical(...):

sufficiently-logical((23.28)
$$s_1 = v_i \geq s_{i,i},$$ (23.29)
$$\varphi_1 = sss_i,$$ (23.30)
$$\phi = v_i \geq s_{i,i},$$ (23.31)
$$\Phi = \Phi_i,$$ (23.32)
$$\models = \models_i)$$ (23.33)

$$[[\Phi \supseteq \{s1,\ldots\} \wedge \ldots \wedge (\Phi \models \varphi_1) \wedge \ldots] \implies (\Phi \vdash \phi)] \implies \Phi \models \phi$$
(23.34)

$$[[\Phi_i \supseteq \{v_i \geq s_{i,i}\} \wedge (\Phi_i \models_i sss_i)] \implies (\Phi_i \vdash v_i \geq s_{i,i})] \implies \Phi_i \models_i v_i \geq s_{i,i}$$
(23.35)

Chipping away at this formula, we observe $v_i \geq s_{i,i} \in \Phi$. Therefore, via the definition of $\mathcal{B}$ notation: $\mathcal{B}(v_i \geq s_{i,i})$. Therefore:

$$= [[\mathcal{B}(v_i \geq s_{i,i}) \wedge (\Phi_i \models_i sss_i)] \implies (\Phi_i \vdash v_i \geq s_{i,i})] \implies \Phi_i \models_i v_i \geq s_{i,i}$$
(23.36)

23.3. PERSONAL IMPLICATION OF THE SUPPOSED SECRET TO SUCCESS

Because every personally provable formula is also a provable formula (i.e. $\Phi \models \varphi \implies \Phi \vdash \varphi$), we can substitute $(\Phi_i \models_i sss_i)$ for $(\Phi_i \vdash_i sss_i)$

$$= [[\mathcal{B}(v_i \geq s_{i,i}) \wedge (\Phi_i \vdash sss_i)] \implies (\Phi_i \vdash v_i \geq s_{i,i})] \implies \Phi_i \models_i v_i \geq s_{i,i} \quad (23.37)$$

Now, if we were to prove the inner conditional $[\mathcal{B}(v_i \geq s_{i,i}) \wedge (\Phi_i \vdash sss_i)] \implies (\Phi_i \vdash v_i \geq s_{i,i})$, then we will have proven $\Phi_i \models_i v_i \geq s_{i,i}$. Let's restate (Condition 2).

$$\Phi_i \models_i sss_i \quad (23.38)$$
$$(23.39)$$

And again recalling that every personally provable formula is also a provable formula, therefore the above expression implies:

$$\Phi_i \vdash sss_i \quad (23.40)$$

Applying the definition of sss_i.

$$\Phi_i \vdash \forall j.[\mathcal{B}_j(v_j \geq x)_{x \leftarrow s_{j,i}} \implies v_j \geq S_{j,i}] \quad (23.41)$$

We instantiate j to i:

$$\Phi_i \vdash [\mathcal{B}_i(v_i \geq x)_{x \leftarrow s_{i,i}} \implies v_i \geq S_{i,i}] \quad (23.42)$$

Players are never confused about their personal success criteria:

$$\Phi_i \vdash sss_i \implies \Phi_i \vdash [\mathcal{B}_i(v_i \geq s_{i,i}) \implies v_i \geq S_{i,i}] \quad (23.43)$$

Applying (Condition 1):

$$\Phi_i \vdash v_i \geq S_{i,i} \quad (23.44)$$

We have proven the inner conditional, and have therefore proven the conclusion:

$$\Phi_i \models_i v_i \geq S_{i,i} \quad (23.45)$$

□

23.4 Interpersonal implication of the supposed secret to success

Lemma 3. *If (Condition 1) Player i deeply believes every player at least shallowly believes they will succeed, (Condition 2) if that player deeply believes every player deeply believes in the supposed secret to success, (Condition 3) if that player deeply believes every player is sufficiently logical to believe in Lemma 2, (Condition 4) if that player deeply believes every Player j's success criteria matches Player i's criteria for Player j (Condition 5) if that player is sufficiently logical to believe in this lemma, sthen (Conclusion) that player will deeply believe every player will succeed.*

Before presenting the proof, let's begin by formalizing the

23.4. INTERPERSONAL IMPLICATION OF THE SUPPOSED SECRET TO SUCCESS

lemma.

$$\text{c-2-1} = \forall j.(\mathcal{B}_j(v_j \geq s_{j,j})) \tag{23.46}$$

$$\text{c-2-2} = \forall j.\Phi_j \models_j sss_j \tag{23.47}$$

$$\text{c-2-3} = \forall j.\textbf{sufficiently-logical}(\tag{23.48}$$

$$s_1 = v_j \geq s_{j,j}, \tag{23.49}$$

$$\varphi_1 = sss_j, \tag{23.50}$$

$$\phi = v_j \geq s_{j,j} \tag{23.51}$$

$$\Phi = \Phi_j \tag{23.52}$$

$$\models \; = \models_j) \tag{23.53}$$

$$\text{c-2-4} = \forall j.(s_{j,j} = s_{j,i}) \tag{23.54}$$

$$\text{c-2-c} = \forall j.\Phi_j \models_j v_j \geq s_{j,i} \tag{23.55}$$

$$\tag{23.56}$$

$$(\text{Condition 2-1}) = \Phi_i \models_i \text{c-2-1} \tag{23.57}$$

$$(\text{Condition 2-2}) = \Phi_i \models_i \text{c-2-2} \tag{23.58}$$

$$(\text{Condition 2-3}) = \Phi_i \models_i \text{c-2-3} \tag{23.59}$$

$$(\text{Condition 2-4}) = \Phi_i \models_i \text{c-2-4} \tag{23.60}$$

$$(\text{Condition 2-5}) = \textbf{sufficiently-logical}(\tag{23.61}$$

$$\varphi_1 = \text{c-2-1} \tag{23.62}$$

$$\varphi_2 = \text{c-2-2} \tag{23.63}$$

$$\varphi_3 = \text{c-2-3} \tag{23.64}$$

$$\varphi_4 = \text{c-2-4} \tag{23.65}$$

$$\phi = \text{c-2-c} \tag{23.66}$$

$$\Phi = \Phi_i \tag{23.67}$$

$$\models \; = \models_i) \tag{23.68}$$

$$(\text{Conclusion}) = \Phi_i \models_i \text{c-2-c} \tag{23.69}$$

$$\tag{23.70}$$

Proof. We begin by assuming all five conditions are true. Next we review the definition of **sufficiently-logical**(...):

sufficiently-logical((23.71)

s_1, (23.72)

s_2, (23.73)

s_3,,... (23.74)

φ_1, (23.75)

φ_2, (23.76)

φ_3,... (23.77)

ϕ, (23.78)

Φ, (23.79)

$\models$) (23.80)

$= [[\Phi \supseteq \{s1, ...\} \wedge ... \wedge (\Phi \models \varphi_1) \wedge ...] \implies (\Phi \vdash \phi)] \implies \Phi \models \phi$ (23.81)

Applying (Condition 5) to the above definition of **sufficiently-logical**(...):

$[[\Phi_i \models_i \text{c-2-1} \wedge \Phi_i \models_i \text{c-2-2} \wedge \Phi_i \models_i \text{c-2-3} \wedge \Phi_i \models_i \text{c-2-4}] \implies (\Phi_i \vdash \text{c-2-c})$ (23.82)

Because every personally provable formula is also a provable formula (i.e. $\Phi \models \varphi \implies \Phi \vdash \varphi$), we can substitute a bunch.

$[[\Phi_i \vdash \text{c-2-1} \wedge \Phi_i \vdash \text{c-2-2} \wedge \Phi_i \vdash \text{c-2-3} \wedge \Phi_i \vdash \text{c-2-4}] \implies (\Phi_i \vdash \text{c-2-c})] \implies$ (23.83)

Now, if we were to prove the inner conditional implies the inner conclusion, we will have proven the conclusion. So, we assume the inner conditional is true.

$\Phi_i \vdash \text{c-2-1}$ (23.84)

$\Phi_i \vdash \text{c-2-2}$ (23.85)

$\Phi_i \vdash \text{c-2-3}$ (23.86)

$\Phi_i \vdash \text{c-2-4}$ (23.87)

23.5. DISCOVERING SUCCESS

Expanding c-2-1 through c-2-4, by recalling the definition of the various components:

$$\Phi_i \vdash \forall j.(\mathcal{B}_j(v_j \geq s_{j,j})) \tag{23.88}$$

$$\Phi_i \vdash \forall j.\Phi_j \models_j sss_j \tag{23.89}$$

$$\Phi_i \vdash \forall j.\text{sufficiently-logical}(\tag{23.90}$$

$$s_1 = v_j \geq s_{j,j}, \tag{23.91}$$

$$\varphi_1 = sss_j, \tag{23.92}$$

$$\phi = v_j \geq s_{j,j} \tag{23.93}$$

$$\Phi = \Phi_j \tag{23.94}$$

$$\models = \models_j) \tag{23.95}$$

$$\Phi_i \vdash= \forall j, k.(s_{k,k} = s_{k,j}) \tag{23.96}$$

$$\tag{23.97}$$

Applying Lemma 2, we have:

$$\Phi_i \vdash \forall j.\Phi_j \models_j v_j \geq s_{j,j} \tag{23.98}$$

Invoking (c-2-4):

$$\Phi_i \vdash \forall j.\Phi_j \models_j v_j \geq s_{j,i} \tag{23.99}$$

We have proven the inner conclusion, and have therefore proven the conclusion:

$$\Phi_i \models_i \forall j.\Phi_j \models_j v_j \geq s_{j,i} \tag{23.100}$$

$\square$

23.5 Discovering success

Lemma 4. *If (Condition 1) Player i deeply believes every player will succeed, (Condition 2) if Player i's success criteria for Player j matches Player j's success criteria for themselves, (Condition 3) if Player i deeply believes their success criteria for Player j matches Player j's success criteria for* themselves, *(Condition 4)* if mutual success is possible, and furthermore if mutual success depends on Player i casting a particular ballot, and *(Condition 5)* if (a) and (b) Player i is sufficiently logical to realize how they should vote, then *(Conclusion)* Player i will cast a ballot consistent with everyone's success.

Proof. Before presenting the proof, let's begin by formalizing the lemma.

$$\text{c-3-1} \not\models_i \forall j. v_j \geq s_{j,i} \tag{23.101}$$

$$\text{c-3-2} \not\models_i \forall j. s_{j,j} = s_{j,i} \tag{23.102}$$

$$\text{c-3-3} = \forall i, j. s_{j,j} = s_{j,i} \tag{23.103}$$

$$\text{c-3-4} = \exists (y_1, \ldots). \{[(b_1 = y_1 \wedge \ldots) \implies (v_1 \geq s_{1,1} \wedge \ldots)] \wedge \\ ([[(b_1 = x_1 \wedge \ldots) \implies (v_1 \geq s_{1,1} \wedge \ldots)] \implies (x_1 = y_1 \wedge \ldots))\} \tag{23.104}$$

$$(Condition1) = \forall i. \Phi_i \text{c-3-1} \tag{23.105}$$

$$(Condition2) = \forall i. \Phi_i \text{c-3-2} \tag{23.106}$$

$$\ldots \tag{23.107}$$

$$(\text{Condition 2-5}) = \textbf{sufficiently-logical}(\tag{23.108}$$

$$\varphi_1 = \text{c-2-1} \tag{23.109}$$

$$\varphi_2 = \text{c-2-2} \tag{23.110}$$

$$\varphi_3 = \text{c-2-3} \tag{23.111}$$

$$\varphi_4 = \text{c-2-4} \tag{23.112}$$

$$\phi = \text{c-2-c} \tag{23.113}$$

$$\Phi = \Phi_i \tag{23.114}$$

$$\models \: = \models_i) \quad \text{c-3-c} = v_1 \geq s_{1,1} \wedge v_2 \geq s_{2,2} \wedge \ldots \tag{23.115}$$

We begin by assuming all five conditions are true. Next we review the definition of **sufficiently-logical**(...):

23.6. MAGICIAN'S RATIONALITY

$$\text{sufficiently-logical}(\tag{23.116}$$
$$s_1, \tag{23.117}$$
$$s_2, \tag{23.118}$$
$$s_3,,\ldots \tag{23.119}$$
$$\varphi_1, \tag{23.120}$$
$$\varphi_2, \tag{23.121}$$
$$\varphi_3,\ldots \tag{23.122}$$
$$\phi, \tag{23.123}$$
$$\Phi, \tag{23.124}$$
$$\models) \tag{23.125}$$
$$= [[\Phi \supseteq \{s1,\ldots\} \wedge \ldots \wedge (\Phi \models \varphi_1) \wedge \ldots] \implies (\Phi \vdash \phi)] \implies \Phi \models \phi \tag{23.126}$$

Applying (Condition 5) to the above definition of **sufficiently-logical**(...):

□

23.6 Magician's rationality

Theorem 3. *If (Condition 1) every player at least shallowly believes they will succeed, (Condition 2) if every player deeply believes in the supposed secret to success, (Condition 3) if every player are sufficiently logical to believe in (a) Lemma 2 and (b) Lemma 3, (Condition 4) if every player deeply believes every player at least shallowly believes they will succeed, (Condition 5) if every player deeply believes every player deeply believes in the supposed secret to success, (Condition 6) If every player deeply believes every player is sufficiently logical to believe in Lemma 2, (Condition 7) If every Player i deeply believes every Player j's success criteria matches Player i's criteria for Player j (Condition 8) if everyone is on the same page about everyone's success criteria, (Condition 9) If success is possible, and furthermore if there exists only one possible solution, (Condition 10) if every player is sufficiently* *logical to realize there exists exactly one possible solution, and (Condition 11) if every player is sufficiently logical to realize what their decision should be then (Conclusion) every player will succeed according to their own criteria for success.*

Proof. Lemma 2 applies to every player via (Conditions 1, 2, & 3-a). Restating the conclusion of Lemma 2 formally, in this situation:

$$\forall i.\Phi_i \models_i v_i \geq s_{i,i} \qquad (23.127)$$

Lemma 3 applies to every player via (Conditions 4, 5, 6, 7, & 3-b). Restating the conclusion of Lemma 3 formally, in this situation:

$$\forall i,j.\Phi_j \models_j s_{j,j} \geq s_{j,i} \qquad (23.128)$$

With Lemma 2 and Lemma 3 applied, we no longer need (Conditions 1 – 7). Thus we show the conclusion of this lemma follows from (Conditions 8 – 10) together with the conclusions from Lemma 2 and Lemma 3.

Moving into the heart of this proof, we formalize (Conditions 8 & 9), starting with (Condition 8):

$$\forall i,j.s_{j,j} = s_{j,i} \qquad (23.129)$$

Towards formalizing (Condition 9), we first express the exists one *or more* solutions:

$$\exists(x_1,x_2,...).(b_1 = x_1 \wedge b_2 = x_2 \wedge ...) \implies (v_1 \geq s_{1,1} \wedge v_2 \geq s_{2,2} \wedge ...) \qquad (23.130)$$

And, since there exists exactly one solution, the following formula expresses (Condition 9):

$$Q = \exists(y_1,...).\forall(x_1,...).[(b_1 = x_1 \wedge ...) \implies (v_1 \geq s_{1,1} \wedge ...)] \implies (x_1 = y_1,...) \qquad (23.131)$$

Via (Condition 10), we assume players are sufficiently logical to deeply believe there exists exactly one solution.

$$\forall j.\Phi_j \models_j Q \qquad (23.132)$$

23.6. MAGICIAN'S RATIONALITY

Via (Condition 11), we assume players are sufficiently logical to figure out the following...

Assume, without a loss of generality, $b_i = C$. □

Part III

Appendix: Zen lemonade garden

Chapter 24

Introduction

Zen lemonade garden is an approach to teaching, and has three chief distinguishing characteristics. First, once a student has learned a lesson, they are then qualified to teach that lesson. Thus, Zen masters are able to distribute their teaching workload to students. Second, students may only be taught by the least qualified teacher available. This creates isolated gardens of teaching, where lower-level students can't pester more advanced teachers—allowing the more advanced students to level up regardless of what's happening at the lower levels. Third, teachers use riddles to teach, instead of relying exclusively upon one-on-one instruction.

24.1 Koans

Koans, enlightening riddles, are nice because they allow a teacher to transmit many lessons quickly, and then the students must study the lessons on their own, on their own time, without relying upon further direct one-on-one instruction.

For example, a teacher might spend an eighth of a work day (1 hour) communicating a personalized lesson to a student, and then the student would spend four work days studying the lesson, before learning the lesson. In this scenario, a teacher could teach eight lessons (eight koans) in one day, and every student would level up after the fourth day.

24.2 Queues

But, what would teachers do if it weren't for koans? Then, I'd imagine teachers would use *queues*—students would need to wait in line for personal instruction from teachers.

At first glance, it might seem it would take 8 days to teach 8 students. However, because students teach each other, the teaching process grows quickly. The details are tedious, but here they are in case you're interested. After the first day, there are now two teachers (because the first student has become a level-1 teacher). But, since students may only be taught by the least qualified teacher available, the Zen master doesn't teach on the second day—only the new teacher (the level-1 teacher) teaches one level-0 student. Thus, on the third day there are two level-1 teachers, and they each teach one student. Thus, on the fourth day there are four level-1 teachers, and four level-0 students. And, finally after the fourth day, the four level-1 teachers teach the four level-0 students, and finally every student has reached level 1.

Even in a scenario where there are 16 students, after the fifth day everyone would level up, equally well for both koans and queues. So, we have established that koans and queues have the same performance when teaching many students. So, what's the point of koans then? Why not just use queues?

24.3 Koans shine when there are many antagonists

In our model, an *antagonist* is a player who wishes to undermine Zen by wasting everyone's time. They request teachings as often as possible, but they never actually learn anything, and thus they never level up, and they never become teachers.

Let's look at an example. Assuming there are 15 antagonists, and one genuine student, with koans it takes 5 steps for the genuine student to solve their koan and level up. But

24.3. KOANS SHINE WHEN THERE ARE MANY ANTAGONISTS

for queues, it depends on where the genuine student happens to land in line. On average, it takes about 8 steps for the genuine student to advance, when using a queue. Koans for the win!

To review, koans and queues seem to perform equally well, supposing there are no antagonists. But, koans shine when there are many antagonists. Now, let's take a somewhat precise look at our model, then we will go over experimental results.

Chapter 25

Model

In our model, a koan k is a riddle that requires k_w dedicated *work units* to solve. Furthermore, each koan must be *transmitted*, and it takes k_t work units to transmit koan k. For every koan, $0 < k_t \leq k_w$ and $k_t \leq 1$.

A *koan generator* is a function $f(k_t) \mapsto k_w$ that maps transmission efforts to work efforts. In our model, we define $f(k_t) = 1 + \log_2(1/k_t)$. This koan generator has some convenient properties. For example, it seems mostly equivalent to the performance of queues, supposing there are no antagonists. Let's look at some concrete examples.

- $f(1) = 1$. If someone spends 1 unit of work to transmit a koan, then it takes 1 unit of work to solve. Since, in our model, students begin solving a koan the moment they being receiving the koan, then $k_t = k_w = 1$ implies that the student has solved the koan by the time they have finished receiving the koan.

- $f(1/2) = 2$. If someone spends 0.5 work units to transmit a koan, then it takes 2 units to solve. Which, in this case, means it takes 1.5 work units to solve after it has been received.

- $f(1/3) \approx 2.58$. If someone spends 0.333 work units to transmit a koan, then it takes 2.58 total units to solve.

- $f(1/4) = 3$. If someone spends 0.25 work units to transmit a koan, then it takes 3 total units to solve.

25.1 Players

A player represents a human in our model. Players can be antagonists, genuine students, and Zen masters. The set P contains all players, and may grow over time, as the *process* proceeds. For every player $p \in P$, the value p_k stores the number of koans that that player has solved. For antagonists, p_k always equals zero. For Zen masters, p_k always equals MaxKoans.

25.2 Students and teachers

At the beginning of each "process step," the process assigns students to teachers. A student is a player who has just finished solving a koan (or hasn't solved any), and needs a fresh, new koan. A student may only be taught by a teacher, if that teacher has solved exactly 1 more koans (with exceptions). Formally, a student $s \in P$ may be taught by a teacher $t \in P$, if $t_k = s_k + 1$, or $t_k = N$, where $N > s_k + 1$, and N is the smallest number such that there exists a player p such that $p_k = N$. In Plain English, students may only be taught by the least advanced teacher possible.

25.3 Player steps

During a *process step*, after the process assigns students to teachers, each player makes a *player step*. During each player step, the player has 1 unit of work available to them. The player divides their unit of work across three different tasks, three different types of work: (1) perhaps receiving a new koan (from a teacher), (2) perhaps transmitting 1 or more koans (to any students), and (3) perhaps working towards solving their current koan.

If the student receives a koan k, then the player must perform at least k_t work units solving that koan, during that step. This represents the player solving the koan as they receive it. Since a student must work to receive a koan, the student has less work remaining for other tasks, when they receive a koan.

For a player step, let w_1 equal k_t for a received koan k (or 0, if the player did not receive a koan), let $w_2 = a_t + b_t + c_t + ...$ for the koans the player transmits, and let w_3 equal the time spent towards solving their koan (minus w_1). The following invariant always holds: $0 \leq w_1 + w_2 + w_3 \leq 1$.

25.4 Assigning students to teachers

Recall, at the beginning of each process step, the process assigns students (players who need a fresh koan) to teachers. Each student is assigned an *eligible* teacher, specifically the one with the smallest classroom size. For any given student, there will be one or more eligible teachers—where eligibility means the teacher has solved the fewest koans, while also having solved more koans than the student.

25.5 Conclusion

That's most of the model. One very important detail we've skipped though, is how teachers should balance their teaching efforts and their personal development efforts (solving their assigned koans). This is the subject of the next chapter.

Chapter 26

Balancing teaching and studying

Suppose you receive one or more students for a particular step. How much time should you dedicate to teaching versus personal development, i.e. solving the koan that had been given to you?

26.1 Rational strategy

If you are "rational," i.e. if you are trying to maximize your personal development, it seems you should not teach at all—because teaching takes away from your personal-development time. In the Zen lemonade garden, "rationality" is therefore a problem, similar to the prisoner's dilemma. Because, if everyone is "rational" then only the Zen master teaches, and everyone advances slowly.

26.2 Superrationality

Perhaps genuine students could cooperate by employing superrationality to overcome the problem of "rationality." Recall from Chapter 17: (1) superrational players assume their partners are also superrational players, and (2) superrational players are classically rational, except *all superrational players are guaranteed to use the same strategy*.

In the Zen lemonade garden, suppose a superrational player's only choice is to choose an x value, where x is the "personal-development proportion" ($0 \leq x \leq 1$). To elaborate, x specifies the proportion of time each player reserves for personal development (i.e. time allocated for solving their own current koan). If we assume x is constant throughout the process, and x is chosen before the process begins, then within this framework, what is the x that maximizes the expected utility for each genuine player?

Well, under certain assumptions, the expected utility of each player is equal to the average utility across all players. Therefore superrational players want to maximize average utility, which means players want to maximize *everyone's* utility.

Across a variety of experimental scenarios, I have discovered that setting $x = 0$ tends to maximize average utility. In plain English, superrational players should dedicate 100% of their time to their students, whenever they have students. Thus, setting $x = 0$ is at least approximately superrational. Perhaps there's a proof to be made here, but for the purposes of this book, we will simply rely on experimental results to guide the discovery of our approximately superrational strategy.

This result is auspicious because Zen, in the real world, teaches students to benefit others. Here, benefitting others also benefits you, under certain assumptions.

26.3 Achieving superrationality

Superrationality assumes all players are superrational. So, let's ask the question: *how can we achieve superrationality?* How can we persuade humans in the garden to behave superrationally? Perhaps promises or enforcement might do the trick. But, in our model, we assume that the first koan, the first lesson, teaches players about superrationality and inspires them to behave superrationally. Thus, by the time they become eligible to teach, they behave superrationally—sacrificing personal development to teach to the max, whenever presented the opportunity.

26.4 Experiments with personal development proportions

Let's see some experimental results. For clarity, let's use the variable name `personal_development` instead of x, to refer to the proportion of time each player reserves for personal development (i.e. time allocated for solving their own current koan)

Unless otherwise stated, each experiment uses `num_steps` = 100. Also, a random number of players are added at the beginning of each process step—between 0 and 2 players, inclusive. We randomize player generation, because steady streams of students tend to produce unusually favorable results. Unless otherwise stated, every player is genuine.

Experiment 1. *personal_development = 0*

At the conclusion of the experiment, average utility = 17.06. To compare, if we were to use queues instead of koans, the average utility is 17.10. Not bad. Since there are no antagonists, we wouldn't expect koans to shine above queues.

Experiment 2.

Let's compare these results across varying values for `personal_development`. From the results of 10 experiments, a value of 0 is the clear winner, and 0.5 comes in second place. Also, 0.5 comes in second place in most of my personal experimentation (not reported here). Note, when `personal_development` = 1.0, then only there is only one teacher—the Zen master.

CHAPTER 26. BALANCING TEACHING AND STUDYING

personal_development	Average utility
0.0	**17.06**
0.1	12.89
0.2	13.83
0.3	**14.35**
0.4	13.87
0.5	**14.36**
0.6	13.27
0.7	13.13
0.8	11.8
0.9	10.26
1.0	7.70

But recall, we're using koans to cope with antagonists, so let's see how koans perform with antagonists, with personal_development = 0.

Experiment 3. *personal_development = 0, probability that a new player is an antagonist is 0.5*

For both koans and queues, the average utility = 1. Oh no, what happened? For both the koan scenario and the queue scenario, after the 100th step, there are 54 hardworking genuine students stuck at level 1, and they are stuck because they keep trying to teach a group of 56 antagonists (at level 0). The antagonists never make any progress, and since the level-1 students are dedicating 100% of their time teaching antagonists, the genuine players become stuck.

Experiment 4. *Like Experiment 3, except personal_development = 0.5.*

In this experiment, the average utility turns out to be 14.0, which is almost as good as if there were no antagonists. We cannot compare this performance to a queue system, because the queue system relies on conveying a complete lesson in one step. I.e., with a queue, teachers cannot split time between teaching and studying, within a given step.

Let's do one more experiment for this section. Let's see what happens if ~90% of players are antagonists.

26.4. EXPERIMENTS WITH PERSONAL DEVELOPMENT PROPORTIONS

Experiment 5. *Like Experiment 4, except probability that a new player is an antagonist is 0.9.*

In this experiment, the average utility turns out to be 13.79. Cool, right? Even when ~90% of players are antagonists, the genuine players still fare well.

Chapter 27

Genuine quitters

So far we have explored hardworking genuine students mixed with antagonists. But what about *genuine quitters*—genuine-hearted players, who really wish to learn Zen, but who give up on koans, repeatedly asking for new koans? Let's do an experiment where all the hardworking genuine players are replaced with genuine quitters, and zero antagonists.

Experiment 6. *personal_development* $= 0$, *and all players are genuine and use the "genuine quitting" strategy*

According to the *genuine quitting* strategy, you should give up on your koan if you don't solve it immediately. Then, request a koan, and give up if you don't solve it in 2 steps. Then, for the next koan, give up if you can't solve it in 4 steps. Then 8 steps, and so on.

In our experiment, the average utility landed at 8.05, which is about half the utility if the players had never given up.

Now let's see what happens when there's antagonists, too.

Experiment 7. *personal_development* $= 0.5$, *genuine players use the "genuine quitting" strategy, and 90% of players are antagonists.*

The average utility landed at 6.91. Not too bad—the antagonists didn't spoil things much.

Now, let's suppose that students learn not to give up after they solve their first koan. Specifically, let's suppose students learn to not give up by learning to believe in the supposed secret to success. Recall from Chapter 1, the supposed secret to success says: *if you believe you will succeed, then you will succeed.* Within the context of our model, when you believe you will succeed you do not give up on a koan, therefore you succeed as quickly as possible. Thus, within the context of our model, the *supposed* secret to success is no longer supposed— it's true.

Experiment 8. *Like Experiment 7, except genuine quitters learn not to quit after they solve their first koan.*

The average utility landed at 13.55, which means learning to not quit about doubles utility. These results are quite similar to the results from Experiment 5, in which 90% of players were antagonistic, and players never gave up.

Since we're teaching students to change their behavior, let's teach them to maximize their teaching efforts once they reach level 2. This seems like a good idea, because `personal_development` = 0.5 is really only useful for dealing with antagonists. Once you're inside the garden, in our model, there are no antagonists and setting `personal_development` = 0.0 offers better utility.

Experiment 9. *Like Experiment 8, except genuine players set `personal_development` = 0.0 once they solve their second koan.*

And, average utility lands at 17.82! This result is pretty much the same as Experiment 1, in which there were no antagonists at all, and players never gave up. Cool, right?

Chapter 28

Sybil attacks

Thus far we have been considering an adversary model where each antagonist can only waste the time of *one single* teacher at each step. This represents a scenario where teachers are savvy enough to dismiss potential students, when those students have already received a koan on that turn, from someone else. Who knows, maybe potential students must present an ID card, and teachers can look up a student's ID in a database, to see if that student has already seen a teacher in that step. Whatever, it doesn't really matter, because now we're going to assume antagonists can don disguises, or whatever, and receive a koan from every eligible teacher at every step.

In computer-security jargon, this type of attack is known as a *sybil attack*.

Let's do an experiment. First, we take Experiment 9, and modify it by refraining from randomly adding new players to the process. Instead, we preload the process with: one Zen master, 100 ordinary antagonists, twenty genuine players at level 1 (or above), and ten genuine players at level 0.

First, we do an experiment *without* sybil antagonists, then an experiment *with* sybil antagonists, and for each experiment we measure how long it takes for the level-0 genuine students to level up.

Recall, the experimental setup for 9: (1) `personal_devel-`

opment $= 0.5$, (2) genuine players use the "genuine quitting" strategy, (3) 90% of players are antagonists, (4) genuine quitters learn not to quit after they solve their first koan, and (5) genuine players set personal_development $= 0.0$ once they solve their second koan.

Experiment 10. *As described above, and without sybil antagonists.*

It takes 7 steps for the level-0 genuine students to solve their first koan.

Experiment 11. *As described above, but with sybil antagonists.*

It takes 15 steps for the level-0 genuine students to solve their first koan, which is about twice as slow. Thus sybil attackers can significantly slow down genuine students.

However, are sybil attacks actually threatening in general scenarios? Let's redo Experiment 9, but with sybil antagonists.

Experiment 12. *Like Experiment 9, except with sybil antagonists*

And, utility lands at 17.73, which is almost as good as the result of 17.82 from Experiment 9! How come the sybil attack didn't do anything? Well, for the most part there was only one single eligible teacher available to antagonists. If you think about it, there might be many sybil attackers, but if there's only one target for their attack, then how would wearing disguises and visiting "multiple" teachers hurt anyone, since there's only one teacher to visit—one threshold guardian?

Therefore, for sybil to make a difference, there needs to be at least several, or perhaps many eligible teachers for the antagonists to attack. In the real world, I don't know if Zen lemonade gardens tend to have solo threshold guardians (or perhaps a small number of threshold guardians), but in the experiments we've looked at, sybil doesn't make much of a difference.

28.1 Conclusion

Our final experiment showed favorable results under unfavorable circumstances. In this experiment, 90% of players were antagonists conducting a sybil attack and 100% of neophytes quit frequently. The Zen lemonade garden achieved favorable performance (comparable to a non-attack, non-quitting scenario) by: (1) relying on koans instead of queues, (2) teaching neophytes not to give up after solving their first koan, (3) having novice teachers divide their time equally between teaching and studying, and (4) having more advanced teachers dedicate 100% of their time to teaching, whenever they had one or more students, and (5) assuming all genuine players would behave superrationally.

Chapter 29

Conclusion

In the Zen lemonade garden, students do not wait in queues to see their teachers. Rather, teachers hand out riddles to their students, *koans*, as quickly as needed—eschewing queues altogether. Under certain mathematical assumptions, koans perform just as well as queues. And, even better, koans still work well when the teaching system is under siege by a large army of frivolous phonies. Even if the phonies don disguises to maximize the amount of time they waste for everyone, the Zen lemonade garden performs almost as well as if there were no antagonists at all.

In the experiments we looked at, for the Zen lemonade garden to succeed, certain conditions must be met. For example, neophytes must quickly learn to believe in themselves, so they can give up on giving up, because giving up is irrational (in this model), unnecessarily slowing themselves down and clogging the system for everyone else.

However, we cannot rely on the classical definition of "rationality" for the Zen lemonade garden to succeed. For everyone to succeed, the students and teachers must cooperate. While it might be "rational" for students to advance more quickly by abstaining from teaching, if everyone is rational, then the system becomes clogged, because no one is teaching. Therefore, for the Zen lemonade garden to succeed, students must cooperate by volunteering their time to teach.

Experimentally, we saw that if teachers teach as much as they can, then everyone, on average, benefits as much as possible. This cooperation is desirable.

If we were to lean on preexisting theories of game theory, we could try to achieve cooperation solely through *superrationality*. Essentially, if everyone agrees to be superrational, beyond traditional "rationality," then everyone will cooperate, and the Zen lemonade garden will perform well. But how do you persuade everyone to cooperate to choose superrationality, so that everyone can cooperate? That's a problem.

But, suppose a Zen lemonade garden is a school of magic. Zen is magical after all. In other words, suppose every teacher believes in the supposed secret to success; suppose every teacher holds as an axiom "if you believe you will succeed, then you will succeed." Then, in this case, we do not need to persuade teachers to cooperate. Rather, a shared belief in the secret to success logically implies *magician's rationality*—helping each other out, and we will succeed.

www.ingramcontent.com/pod-product-compliance
Lightning Source LLC
Chambersburg PA
CBHW060848170526
45158CB00001B/280